DAMON GAMEAU

THIS BOOK WILL CHANGE THE WAY YOU THINK
ABOUT 'HEALTHY' FOOD

MACMILLAN

CONTENTS

FOREWORD
BY DAVID GILLESPIE

Ten years ago I lost 40 kilograms by doing just one thing: I stopped eating sugar. I did it because the research I was reading led me to the undeniable conclusion that my sugar consumption was keeping me fat. But I didn't stop reading when the weight started coming off. I discovered more and more research on the damage being caused by sugar.

Sure, sugar was keeping me fat. But that was the least of its charms. The research was clear: sugar also destroys our teeth, our gut lining, our liver, our pancreas, our kidneys and our heart. And the damage didn't stop there; it was increasingly clear that sugar could destroy our state of mind as well. The research tells us that consuming sugar, even 'in moderation' (a phrase invented by the food industry to give you licence to eat as much as you like), is likely to be the cause of the modern epidemic of anxiety and depression.

As I was trying to fit into a smaller pair of jeans, Damon was independently figuring out that he was a better person off sugar. He was, and is, a fit and healthy young man, so he wasn't looking to sugar as a weight loss tool. But he was curious and interested in how sugar affected his moods. So he decided to see if reducing sugar made any difference.

He dropped the white stuff and almost immediately experienced a dramatically improved sense of wellbeing. But was it the sugar? Or just coincidence? Damon was in a very good place in his life at the time. He was (and is) immersed in a happy relationship, looking forward to the birth of a delightful heir and was starting to hit some home runs in his career as well. His acting career was flourishing and his first film as a director had just run away with an against-the-odds win at Tropfest in 2011.

How much of his improved sense of wellbeing was because he removed as much sugar as he could from his diet? Or, was it down to the fact he was in a good place anyway? Was it the chicken or the (Easter) egg that came first?

At some point, Damon heard me talking about sugar consumption and health on the radio. He asked if we could meet. I am not a creative person by nature, and have not met many professional actors (well, none, unless you count politicians), but was I in for a treat. Damon was an earnest blur of energy and ideas. He wanted to find out whether the absence of sugar could take the credit for how he felt. He wanted to do an experiment on himself. He wanted to eat the Australian daily average sugar consumption for two straight months and see what happened. And he wanted to film it. Oh, and he wanted the subsequent movie to make a glorious and spectacular impression, no matter what the result.

I told him he was a sandwich short of a picnic. If even half of the research I'd read was true, he was likely putting himself at risk of significant harm. He reckoned there was a good chance I was right (about the harm not the sandwich!) but, he reasoned, what better way to find out? And if I was right, what better way to show millions of people? As I was soon to discover in spades, when Damon has a plan, your choices are follow (and enjoy the ride) or get out of the way. I chose to follow.

Damon took a risk for us all. He did it for you, your kids, those you hold dear. And, like most curious people, he did not think of the consequences. And there were consequences, not just for Damon, but for his, by then, pregnant partner Zoe (whose strong and unwavering support underpins every frame).

What you have in your hands is the story of a man who wanted to answer a question for himself, fully aware that it would help others. Damon stepped up. He couldn't help himself. What he has proven in real time, with the cameras rolling, would never get research approval (I warned him of that). But his careful, intelligent planning means that you now know the devastating effect sugar will have on you – is having on you. The people creating your food, providing your food, creating health policy about your food, have nowhere left to hide in the face of such compelling evidence.

Damon (an ordinary, extraordinary bloke) and his family have taught me a lot, and I have gained a friend on the way. He's a humble, practical, wicked fella who just wants you to know the truth. Read on and prepare to be inspired.

DAVID GILLESPIE HAS AUTHORED NUMEROUS BOOKS, INCLUDING THE BEST-SELLING *SWEET POISON* BOOKS ON THE DANGERS OF SUGAR.

INTRODUCTION

Apart from climate change, or perhaps who the best One Direction singer is, there is not a more fiercely debated, opinionated or instagrammed topic on the planet than nutrition. Trying to find answers, especially if you're looking online, can feel like you're wading through a mass of confusion and contradictions fuelled by inscrutable scientific data, dogmatic culinary gurus and pages of vitriolic comment. It is now unclear whether bread is a friend or an enemy; dairy is scary but butter is better; and what about feta? Some foods have even stepped into some kind of nutritional phone-booth and emerged as 'super foods'.

Be it health advocates, the media, food bloggers, friends – it seems everyone has a different opinion on what we should and should not eat.

Amid all the debate though, there is one ingredient that seems to be the current darling of the nutritional paparazzi: sugar.

Over the past few years, sugar has dominated the headlines, with plenty of speculation and bickering over its effects on our health. I felt quite ambivalent about it until two blue lines appeared on my better half's pregnancy test. The topic then seemed to rocket in importance so I decided I better bypass opinion and find out for myself.

We can read all we like about a subject but in the end all we ever have is our own experience. This book is about sharing my sugar experience.

60 DAYS, 40 TEASPOONS

For 60 days I would test out a high-sugar diet by consuming 40 teaspoons of sugar a day. 'You're a lunatic!' I hear you cry – but the scary fact is that 40 teaspoons of sugar is what many Australians are consuming every day (with Britons consuming on average 34 teaspoons a day), and teenagers are having even more.

There would be a twist, however. Unlike the average teenager, I would only eat perceived 'healthy' foods – foods generally believed to be low in, or free from, sugar. This meant I had to reach my 40 teaspoons a day without soft drink, chocolate, ice-cream or confectionery. Instead, my diet would be made up of foods like low-fat yoghurt, cereals, muesli bars, juices and sports drinks: all of which are full of 'hidden sugars'.

During the experiment I was monitored by a team of doctors and scientists. I also embarked on a sugar-fuelled adventure across Australia and the USA, filming the whole thing for a feature documentary film. I met some wonderful characters, interviewed the world's leading experts on sugar, watched my mind and body change and learnt more about this little white substance than I ever thought I would.

This book is divided into four parts. For those of you who like a good story (or a bad one depending on your taste), the first section follows the ups and downs of my sugar-eating lunacy. The second part is all about the science: the facts come complete with pretty pictures to help explain what sugar did to my mind and body. The third part describes how I successfully removed sugar from my diet. It also provides some simple assistance for those who may also want to reduce their sugar intake and live a healthier life. The fourth part includes an all access pass to recipes that helped me during the post-experiment detox.

I went into this adventure without knowing what to expect. Some of what I learnt was pretty frightening – but it has changed my life and subsequently the life of my child. My hope is that in some small way it might change yours too.

'SUGAR ISN'T EVIL, BUT LIFE IS JUST SO MUCH BETTER WHEN YOU GET RID OF IT.'
KATHLEEN DESMAISONS,
AUTHOR OF *THE SUGAR ADDICT'S TOTAL RECOVERY PROGRAM*

THE EXPERIMENT:
60 DAYS OF SUGAR

PART ONE

> 'WE HAVE ENTERED INTO A FAUSTIAN PACT WITH SUGAR; IT'S A SHORT-TERM REWARD FOR A LONG-TERM TRAGEDY.'
> DAVID WOLFE, NUTRITIONIST

MY STORY

When I was a baby in the late 1970s, a certain blackcurrant drink was extremely popular. According to the television adverts, it was full of vitamin C, essential for raising strong, healthy children. My dear mum, believing this nutritional advice – which was also promoted by the Nursing Mothers' Association at the time – poured it down my gob like coolant into a boiling radiator. Now, this blackcurrant drink may have contained 'hints' of vitamin C, but it contained 'large kicks under the table' of sugar. And so, at the tender age of four, I had to make a very uncomfortable trip to the dentist to have five baby teeth extracted.

But then, like some wronged action hero, my adult teeth arrived and immediately sought revenge. It was almost as though they were expecting to be flooded with non-stop blackcurrant drink and so they beefed up, injected extra enamel and plunged from my gums like two giant sheets of ice breaking from a melting polar cap. I was a rare boy whose face grew into his teeth. Sniggering school friends would suggest I project home movies onto them and asked whether the tooth fairy claimed workers' compensation for damage caused as she carried them off into the night with her little wings.

Despite the taunts, my walrus chompers and I continued to carve through a whole range of sugary foods. For the next 20 years (with the help of some sensational dental work), I had not one 'sweet tooth' but two giant 'sweet teeth' leading the way (at night because the light would bounce off them). I drank a can or two of Coke a day (I especially loved the vanilla-flavoured one). I smashed chocolate biscuits by the packet, gulped down sweet yoghurt drinks, heaped sugar onto my Weetabix and demolished an unknowable number of Peanut Chews chocolate bars at the school tuckshop. Now I don't for a second want this to read as horrible parenting. This was a time when all the headlines were about the dangers of fat and the low-fat movement was in full swing. The sugar train was fully loaded, travelling at high speed and bounding freely down the mountain towards the unsuspecting villagers.

A NEW DIRECTION

Then, in 2002, I had the great pleasure of working on a film called *The Tracker*, directed by Rolf de Heer and starring one of Australia's true national treasures, the Aboriginal actor David Gulpilil. David and I became good friends and after filming he invited me to spend two weeks with his family in his hometown of Ramingining, Arnhem Land. To this day, I have never been to a more foreign land – and I was in my own country.

There were many moments on that trip that shaped my then confused self-conscious twenty-something self, but I have two clear and lasting memories: first, the warmth and generosity of the people I met, and secondly, the enormous consumption of a well-known black fizzy liquid.

I remember watching many of the Aboriginal women pouring Coca-Cola into their babies' bottles and feeding it to their crying children. When I questioned this, an elder woman marched me to the only store in town, pointed to a poster featuring young, beaming models with huge teeth to rival mine and said simply: 'Happy juice.' Her honesty of interpretation of that advertising has always stayed with me.

The years passed and, as often happens with a man just before it's too late, he meets a sensational woman. That woman will iron out a few creases, tame his wandering eye and give him a gentle reminder that he is, in fact, an adult. For me, it happened at 32. My body and my behaviour reflected years of sugar abuse. Then along came Zoe, a vision of beauty, balance and vitality, whose skin radiated as much as her personality. She understood that the food we put into our bodies is instrumental to the way we appear, feel and view the world.

In the formative weeks of a relationship, men will pretend to be interested in a whole range of things to win the affections of their new love. For me, it was pretending to enjoy cucumber and kale smoothies, quinoa salads with chia seeds, and plain yoghurt with berries. Love really does work miracles though, and as I got used to this new way of eating – and, crucially, as I began to feel better in both body and mind – I discovered I actually enjoyed it. As our relationship developed, I naturally ate less and less refined sugar, until one day a health-conscious friend of mine suggested I just remove it altogether. Truth be told, the thought made me a little uncomfortable but I decided to give it a go.

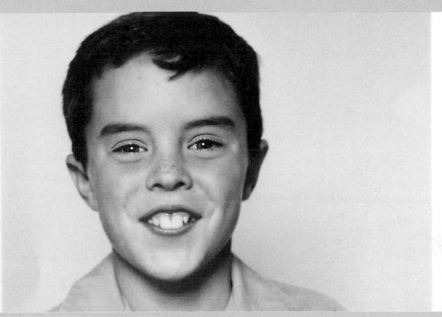

I was a rare boy whose face grew into his teeth.

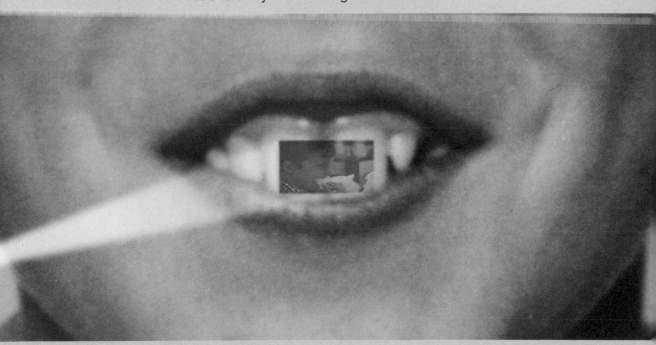

Sniggering friends would suggest I project home movies on to them.

'IF THE AVERAGE AUSTRALIAN FAMILY OF FOUR HAD TO BUY THE AMOUNT OF SUGAR THEY ARE CONSUMING, THEY WOULD BE GOING TO THE SUPERMARKET, TAKING SIX ONE-KILOGRAM BAGS OF SUGAR OFF THE SHELF – SIX; PUTTING THEM IN THE TROLLEY, TAKING THEM HOME, EATING THEM ALL THAT WEEK, THEN GOING BACK THE NEXT WEEK AND DOING IT ALL AGAIN.'

DAVID GILLESPIE, AUTHOR OF *SWEET POISON*

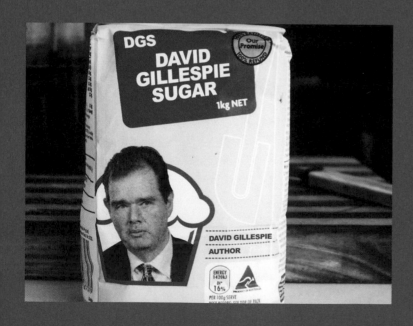

REMOVING SUGAR

The first thing I noticed when I stopped eating sugar was how much I craved sugar. I realised that, despite my much improved diet, sugar still had a strong hold on me. I began to read food labels and was surprised at how many seemingly healthy foods contained sugar. I now understood that the piece of chocolate I 'allowed' myself at the end of each day was really only the tip of the iceberg – an iceberg made of sugar.

What surprised me the most was how good I started to feel. My mood swings felt more 'playground' than 'theme park' and I felt lighter and more present in the world. And then there was the vanity. Being an ego-driven actor at the time, I basked in the flood of comments about my glowing skin, my radiant blue eyes and, I quote, 'sparkly demeanour'. Shallow maybe, but that's all it took for me to finally ditch the Choc Wedge and switch to the veg.

A couple of years later, I won Tropfest, a short-film competition, and was subsequently given the opportunity to make a feature film. At the time, I was writing a show about what the effects may be if you gave a healthy person nothing but hospital food for a month while they lay in bed reading trashy magazines and watching daytime television. Then one day my friend Charlie told me about fructose, and how theories were emerging about the damage it caused. Later, I heard the author David Gillespie talking about sugar on the radio, and the germ of an idea began to form. One thing led to another, plans were floated and then sunk; people and friends chimed in with thoughts and suggestions. Before I knew it, I was once again staring at a bowl of sugary cereal – only this time a camera crew was there to capture every mouthful.

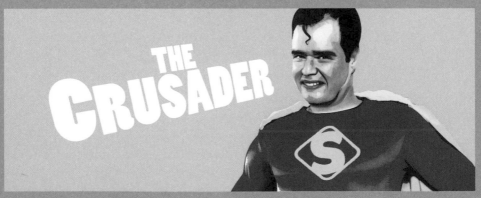

THE SUGAR SQUAD:
Dr Debbie Herbst, Dr Ken Sikaris, Sharon Johnston, David Gillespie

'I EAT A LITTLE PIECE OF CHOCOLATE
EVERY DAY BUT I UNDERSTAND THE
ADVERSE EFFECTS IF I EAT TOO MUCH OF
IT.'

PROFESSOR BARRY POPKIN, UNIVERSITY OF NORTH CAROLINA

THE EXPERIMENT

The first step in the experiment was to assemble a cracking team that could guide me along the journey and monitor any changes to my body. This was to make sure that A: things were done correctly and B: I didn't die – although what a film that would have made. I enlisted a local GP (to monitor my overall health), a clinical pathologist (to check my blood readings), a nutritionist (to guide my food choices), a sport scientist (to accurately measure my weight, and also provide the coolness factor) and David Gillespie, author of the best-selling book *Sweet Poison* (to help me read food labels, understand the science, and perhaps acquire some sugar 'street cred').

At the start of the experiment, I hadn't touched any refined sugar for nearly three years, and I hadn't consumed alcohol or caffeine for five years. I am clearly in love with this girl, aren't I? I also just became addicted to feeling clear in my mind. (Perhaps I am a clarity junkie? I just visualised hooking up with a 'meditation dealer' on a wild night in Kings Cross and 'getting present' together.)

All the tests pre-experiment showed that I was in good health, especially in my liver, which often bears the brunt of an excessive lifestyle. My assembled team agreed I had a great body on which to conduct the sugar experiment – thanks, guys – because it was not affected by anything that would cloud or confuse the results, such as caffeine, prescription medications, drugs or alcohol (this test would have been a disaster in my early twenties).

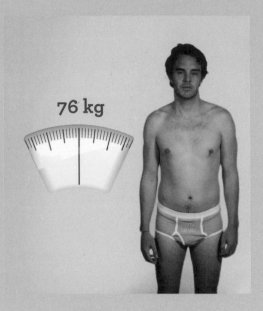

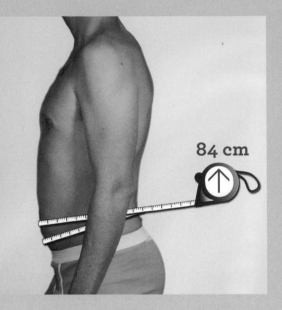

Because I hadn't eaten much sugar for nearly three years and hadn't drunk alcohol for five, my body was the perfect guinea pig to really see the effects of sugar.

For the next 60 days, things would be very different. Not only would I eat sugar – and lots of it – I would also remove a lot of the fats from my diet (found in avocado, cheese, coconut products, etc.). Everyone in the team was looking forward to seeing what would happen to my body, in an excited – and slightly sadistic-masochistic way. We really didn't know what to expect.

IT'S IMPORTANT TO KNOW THAT PRE-EXPERIMENT,
MY DAILY DIET LOOKED SOMETHING LIKE THIS
(PLEASE NOTE THE INFLUENCE OF A GOOD WOMAN):

BREAKFAST:
three poached or scrambled eggs with avocado,
spinach and bacon (no toast)

LUNCH:
salad with kale/lettuce, peppers, tomato,
cheese and tuna

DINNER:
steak or chicken cooked in butter or coconut oil
with lots of green vegetables

SNACKS:
nuts (like almonds);
hummus with carrot and cucumber sticks;
Zoe's homemade pâté with carrot sticks.

THE RULES

→ I must consume 40 TEASPOONS OF SUGAR A DAY.

→ These must be 'hidden' sugars found in 'HEALTHY' foods and drinks, such as breakfast cereals, muesli bars and juices. So NO SOFT DRINKS, ICE-CREAM, CONFECTIONERY or CHOCOLATE.

→ The sugar must consist of SUCROSE and FRUCTOSE specifically, whether they are 'added' or naturally occurring. Despite carbohydrates like bread breaking down into a type of sugar in the body, they will not be counted. (See pages 88–89 for more info about the different types of sugar.)

→ I must always CHOOSE LOW-FAT foods.

→ I must MAINTAIN MY EXERCISE routine: three laps of my long, steep garden twice a week and ten minutes in my homemade gym.

MY HOMEMADE GYM (This consists of dumb-bells made of two 20-litre water bottles and a tent pole. In your face, Fitness First!)

THE SORTS OF FOODS I WAS EATING
BEFORE THE EXPERIMENT

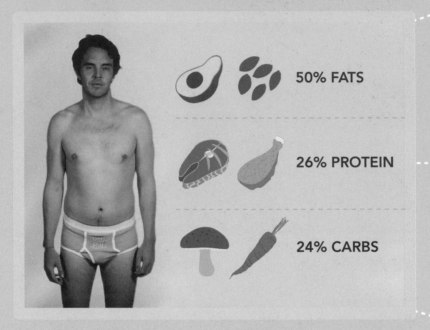

50% FATS

26% PROTEIN

24% CARBS

Total calorie intake is approximately 2,300 (9,600 kJ) per day with roughly 50 per cent coming from healthy fats (like avocado and nuts), 26 per cent from protein (eggs, meat or fish) and 24 per cent from carbohydrates (fresh vegetables). As you can see there is no refined sugar.

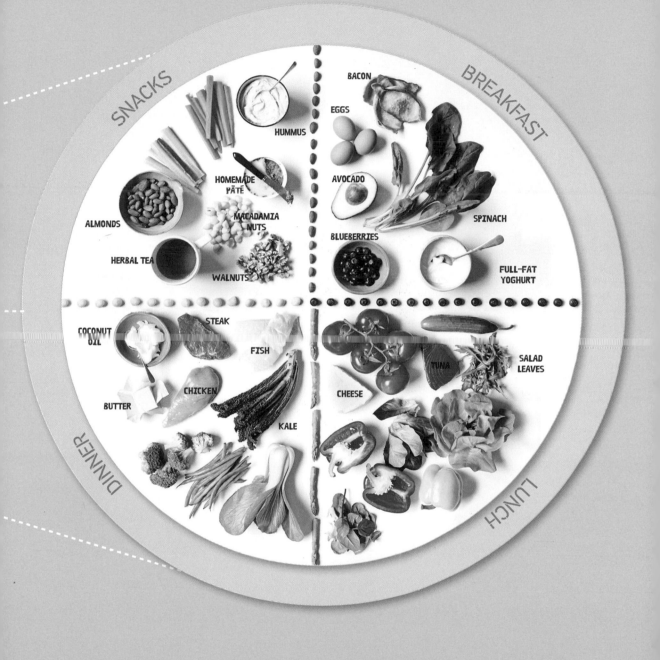

SNACKS

HUMMUS

HOMEMADE PÂTÉ

MACADAMIA NUTS

ALMONDS

HERBAL TEA

WALNUTS

BREAKFAST

BACON

EGGS

AVOCADO

SPINACH

BLUEBERRIES

FULL-FAT YOGHURT

DINNER

COCONUT OIL

STEAK

FISH

CHICKEN

BUTTER

KALE

LUNCH

TUNA

SALAD LEAVES

CHEESE

THE SORTS OF FOODS I WILL BE EATING
DURING THE EXPERIMENT

Everybody in the team was looking forward to seeing
what would happen to my body in an excited, and slightly
sadistic-masochistic way.

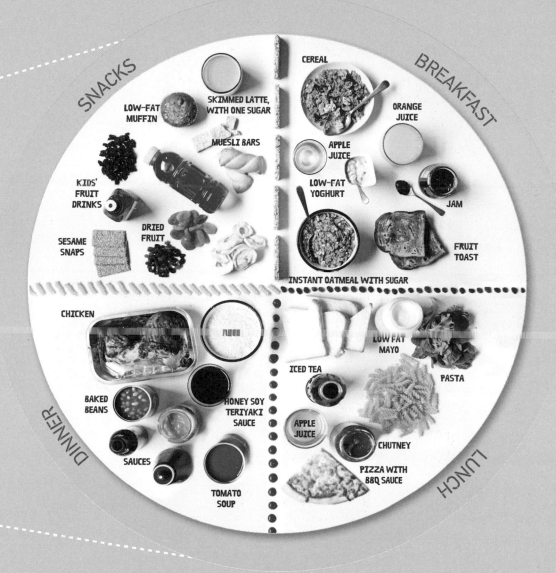

SNACKS

LOW-FAT
MUFFIN

SKIMMED LATTE,
WITH ONE SUGAR

MUESLI BARS

KIDS'
FRUIT
DRINKS

SESAME
SNAPS

DRIED
FRUIT

BREAKFAST

CEREAL

ORANGE
JUICE

APPLE
JUICE

LOW-FAT
YOGHURT

JAM

FRUIT
TOAST

INSTANT OATMEAL WITH SUGAR

CHICKEN

BAKED
BEANS

HONEY SOY
TERIYAKI
SAUCE

SAUCES

TOMATO
SOUP

DINNER

ICED TEA

LOW FAT
MAYO

PASTA

APPLE
JUICE

CHUTNEY

PIZZA WITH
BBQ SAUCE

LUNCH

The sugar I eat must be 'hidden sugars' found in foods that many people would consider healthy.

IT BEGINS: BOOSTER ROCKETS FOR BREAKFAST

My first meal was breakfast. I chose a bowl of Just Right cereal, two scoops of low-fat yoghurt and a 400 ml glass of apple juice. This apparently healthy meal provided 20 teaspoons of sugar – all before I had even reached the gossip section of the newspaper. That's more than twice the amount the American Heart Association recommends men consume in a whole day (I would use the Australian teaspoon recommendations but we don't have any).

After not having had refined sugar for two years, the first surprise was the immediate effect on my mood. I had naïvely assumed the first few days would be quite fun. I was secretly looking forward to having a few foods and drinks I hadn't tasted in ages. Schmuck.

Straight after my Just Right breakfast–dessert combo, I noticed the sudden ramping up of my adrenals and the booster rocket–type surge to my overall energy. It was like taking a set of jumper leads to my chest. I flicked into hyper-drive and casually threw a wave to the Millennium Falcon as I sped by.

I then looked for a target. I needed to talk or more accurately offload. Poor Zoe copped a spray of vowel bullets from my mouth gun. I could see the reality of the next 60 days descend on her in one terrifying swoop. She is a smart woman and that look of fear also implied a deep knowing that what goes up must come down. Sure enough, 45 minutes later life-of-the-party, what-are-we-doing-next Damon had morphed into grumpy, edgy, who-gives-a-rat's-arse Damon.

Now I acknowledge I might have been particularly susceptible to the sugary high (and low) because I hadn't had any in a long time, but I was still shocked, and so was Zoe. In that moment we knew that for the next 60 days there would be another woman in my life. She wore a white coat made of tiny crystals and would definitely be the one wearing the pants in the relationship.

HIDDEN SUGAR HOTSPOTS

Some 'healthy' cereals contain nearly as much sugar as the 'bad' ones you avoid. I began to notice the health claims on the cereal boxes too: 'Contains Essential Vitamins'; 'Great For Energy'; 'Wholegrain Goodness'. (How about 'Crammed With Sugar' or 'Superfluous Slogan'?)

LOW-FAT YOGHURTS are the secret dairy dens of the underground sugar gangs. Although they are marketed as a 'healthy' breakfast option, they should be swiftly moved to the dessert section.

In fact, low-fat foods in general are often higher in sugar than the full-fat variety. I noticed this with some weight-loss meal plans too.

Some MUESLI BARS or 'health bars' contain more sugar than Kit Kats or Milky Ways.

DRINKS are by far the swiftest delivery system for sugar. Juices, vitamin waters, breakfast drinks, sports drinks, organic sparkling juices, flavoured milks and **SMOOTHIES** are all high in sugar.

SOME JUICES (including apple or cranberry) contain more sugar than Coca-Cola in an equivalent-sized glass.

LOW-FAT MAYONNAISE and salad dressings are sugar hotspots.

PASTA SAUCES (and sauces in general) have sugar lurking in them – check out tomato and barbecue sauce for a laugh – or a wince.

Sugar loves CANNED SOUPS.

KIDS' LUNCHBOX SNACKS are a real concern. Fruit bars, fruit snacks, fruit rolls, jellies, sesame snaps and yoghurt muesli bars are all outstanding sugar vehicles.

When it comes to packaged food, look out for the words 'natural', 'mother nature' or 'valley' and/or a picture of a bee, flowers, a rolling meadow or a bright sun. These labels are worth a closer read. They will often say 'evaporated cane juice', 'organic palm sugar' or 'fruit juice concentrate' instead of just cutting to the chase and saying 'sugar'. (See page 138 for more about other terms used for sugar.)

Many items in health food shops appear healthy but can still be packed with sugar or agave. It really doesn't matter if the sugar is organic or not, the body still deals with it in the same way. Agave is often 90 per cent fructose. (See page 88 for more about fructose.)

I was pretty taken aback by these revelations. How many of us think that if we stop adding sugar to our coffee or tea and avoid sweets and chocolate, we no longer eat sugar? Wakey, wakey, good citizens of earth, we are practically bathing in the stuff.

5   8.5   13   5   15   9.5   9.5   5.5   7.2

■ = 1 TEASPOON OF SUGAR

SUGAR COMPARISON DAY:

THE DAY I ATE 40 ACTUAL TEASPOONS OF SUGAR

The first week had been quite an eye-opener into how prevalent sugar's role was in our food supply. So prevalent, that it is now found in around 80 per cent of all items available in a supermarket. I wanted to find a way to help people clearly understand this and to see what 40 teaspoons of sugar looked like stripped of its attractive and cleverly marketed packaging. Here is an excerpt from a diary I kept through the experiment:

FROM THE DIARY: Today was a first for me – and I assume a first for the many people who witnessed it. Today was about making a point that encapsulated all that I have learnt so far on my journey down the candied rabbit hole.

At breakfast, I considered a bowl of Just Right cereal with a scoop of low-fat yoghurt. I then took a bowl of sugar-free Vita Brits, spooned 10 teaspoons of sugar on top – the same amount of sugar found in the Just Right and yoghurt combo – and proceeded to eat the lot.

I have never had that much actual sugar in my mouth before. I swear I heard my dentist scream (with joy at the incoming cash flow) as the sugar burrowed its way in between my vulnerable molars and incisors.

A rapid eye-twitch almost instantly engaged and I thought long and hard about continuing with the plan. However, stupidity and 'suffering for my art' prevailed.

Into town I went. My mid-morning snack was supposed to be a cup of 'healthy' frozen yoghurt. However, instead I bought a tub of sugar-free Greek yoghurt and heaped 11 teaspoons of sugar on top. As I tucked into my snack next to a busy city footpath, people stopped and stared and accused me of being 'mad', 'crazy' and 'foolish'. Correct on all accounts. I am learning these three succinct words may perfectly describe the current food model in our society.

And so to the food court for lunch and a beautifully cooked half-chicken. I was about to pour half a pack of teriyaki sauce onto my chook, but instead I opted for the 4 teaspoons of

'THE SIMPLE MESSAGE IS THAT THE MORE PROCESSED FOODS YOU EAT, THE MORE FRUCTOSE YOU HAVE ADDED TO THE FOODS.'
JEAN-MARC SCHWARZ, TOURO UNIVERSITY, CALIFORNIA

sugar it contained. I may have invented the world's first chicken dessert. Cue the collective lunch-crowd jaw-drop. (They scoff now, but one day they will all be buying my 'Gameau Farm's Chicken Dessert'.)

I washed the sugary chicken down with 8 teaspoons of sugar in a warm glass of water, the equivalent of a bottle of Powerade. My nervous system immediately switched to light speed and once again Han Solo took over the controls. I sat there for a while and took in this new and intriguing state of consciousness. Colours seemed brighter and I felt like I could hear everyone's conversations at the same time, which in an inner-city food court is quite terrifying.

Fifteen minutes later, I was passed out on a huge couch in the middle of the shopping centre. A security guard woke me, without even offering a cuddle.

I wandered around for a while and then at 3 pm I watched all the office workers scurry out onto the streets to inhale a cigarette or a can of Red Bull. Perfect, I thought. It's time for my afternoon pick-me-up.

I looked longingly at a 'superfood' apple and cranberry ripple bar and then promptly produced two water crackers, wedged seven sugar cubes between them and bit down hard. It made a magnificent crunching noise but those crackers are designed for delicate French cheeses and pesto dips; sugar really has no place being anywhere near them.

At 7:43 pm I threw up, which I was quite happy about. I had a salad and a cup of tea with Zoe and am now good as gold. Night.

BREAKFAST
Just Right with low-fat yoghurt

= **10** tsp

MID-MORNING SNACK
Frozen yoghurt

= **11** tsp

40 tsp

'I may have invented the world's first chicken dessert. They scoff now, but one day they will all be buying my "Gameau Farm's Chicken Dessert"!'

LUNCH
Teriyaki chicken and Powerade

= **4+8** tsp

AFTERNOON PICK-ME-UP
A Go Natural apple
and cranberry ripple bar

= **7** tsp

A GLASS OF JUICE IS NOT
THE EQUIVALENT OF FOUR APPLES,
IT IS THE EQUIVALENT TO
THE SUGAR OF FOUR APPLES.

MY JUICEPIPHANY
AND THE FIBRE FACTOR

It became evident very quickly that all this sugar-eating was kicking my rapidly expanding arse. The 'Nobel Prize for Tolerance' should be invented and then awarded to Zoe. She got a glimpse of what I may have been like as a four-year-old jacked up on the blackcurrant drink. But somehow on Day 12 a moment of clarity burst through the petulance . . .

FROM THE DIARY: Today I took four shiny red apples and contemplated eating all of them. There was no way I could have got through even two of them. Then I pulled out the juicer from the cupboard and fed those succulent fruits into the whirring metallic jaws of all that embodies man's convenience. It was here my 'juicepiphany' occurred.

I realised we have developed such ravenous cravings for a sweet hit that we have built a kitchen appliance that strips the fruit of its precious fibre and yields only the sugary nectar within. Mother Nature gave us fruit in a perfect package with fibre that tells us when we are full, and the perfect amount of sugar for our finely tuned bodies to cope with. You won't find juicers in the Serengeti or in the heart of the Amazon!

The four handsome apples I had held in my hand were now obliterated. All that remained was a full glass of the fruit's sweetness. In that glass were 16 teaspoons of sugar and I was able to consume them effortlessly. Deal with that, body! And we recommend this to our children! We push it as a healthy option. It's a madhouse of sugary indulgence.

On this day it occurred to me that much of my sugar intake was coming from liquids and that it rapidly flooded my system. Be it a juice, vitamin water or sports drink, these beverages were a sugary bullet train arriving at my internal organ stations with Japanese rail efficiency. (That is my final train metaphor.)

Let's get one thing clear:

FRUIT AND FRUIT JUICE ARE TWO VERY DIFFERENT THINGS.

Fruit is a good thing if eaten with care. In its natural, carefully constructed casing, fruit is a balance of fibre, nutrients and the right amount of sugars designed to metabolise at the correct speed for our bodies. Many scientists and nutritionists I met during my experiment said that fruit is nature's dessert and when treated that way it is perfectly fine. Most recommended no more than two pieces a day and suggested going easy on the grapes, apples and pears because they are all high in fructose. Just remember that in nature fruits are seasonal. Sweet fruits were not designed to be readily available at any given moment, as they are now. (And real, organically grown fruit is very different from the enhanced versions displayed on supermarket shelves. Go to an organic shop and look at how tiny and withered some fruits can actually look. This is how they should really look, without any 'plastic surgery'!)

Fruit juice, however, gathered a very different response from the people I met. The origins of juicing can be traced back to the 1950s, when Florida orange farmers cleverly transformed their storm-damaged crops into juice to prevent waste. A new craze soon hit the streets, and 70 years later, it shows no sign of disappearing. Juice bars have popped up all over the world. Even packaged juice continues to be perceived as a healthy option. In 2009, the world's population drank 106 billion litres of fruit drinks.

Professor Barry Popkin (I would call him 'Soda Popkin' if we were friends) from the University of North Carolina is a great expert in this field and is a crusader for public health all over the world. He told me: '98 per cent of juices today are a mixture of fruit juice concentrate, water and flavour.' He believes they are having disastrous effects and consuming them is no different from drinking soft drinks.

Jean-Marc Schwarz of Touro University, California, describes our juice and sweetened beverage consumption as a 'tsunami of fructose that floods the liver. It is a big wave of sugar that brings with it some serious consequences.'

Just put yourself in your liver's shoes for a moment. If you saw gently chewed, broken-down pieces of nice shiny apple coming at you, you would have lots of time to decide where everything goes, in a neat, orderly fashion. But if that same apple is thundering towards you as a giant tsunami of juice, then you may start to feel a little rushed in your decision-making.

This removal of fibre also has another consequence. In those early weeks of my experiment, I noticed that because I had removed the fat and fibre from my diet, I never really felt full; no meal satiated me. I learnt just how important fibre is in slowing down the metabolism of what we eat and keeping our system in balance. Unfortunately, we have removed much of it from our diets through the processing of our foods.

(This was also affecting my wallet. I observed that I was spending the same amount on the sugary foods as my pre-experiment diet. I needed to buy more of these foods because I wasn't feeling full from the fat or fibre.)

It's important to understand that dried fruits are also high in fructose. I counted the number of sultanas in a small box – 91. I then tried to eat 91 grapes (the equivalent amount of fructose).

I only got to 24 grapes before the fibre did its job and told my brain that I was full.

91 GRAPES

↓

91 SULTANAS

FRUCTOSE FACT:
THE HUMMINGBIRD IS THE ONLY
ANIMAL THAT CAN USE FRUCTOSE
DIRECTLY FOR ENERGY. IT FLAPS
ITS WINGS AROUND 50 TIMES
PER SECOND.

THE FIRST BLOOD TEST RESULTS

After just three short minutes in a Melbourne laboratory, one of Australia's leading clinical pathologists told me: 'Your liver cells are dying. The changes you have experienced are beyond the day to day. They are very real changes.' And most confronting of all: 'I have never seen fatty liver develop in such a short time, your liver went from the best percentage of men to the worst in just three weeks.'

He also said, 'Your testosterone levels have dropped,' and I said, 'No they haven't,' and punched him in the face. (Not true, he is one of the kindest men you will ever meet. My testosterone had dropped significantly though: yet another Nobel Prize for Tolerance for Zoe.)

This was the most significant day of the experiment so far. It revealed for the first time that sugar was doing something unique to my body.

I will never be able to properly convey the tone of my producer's voice when I told him the results. This one blood test and my now-fatty liver meant we actually had a story: he hadn't just blown a large wad of cash filming some idiot pouring sugar onto his chicken. It's a strange feeling to have someone be delighted when you inform them your liver has fallen apart.

MY FRUCTOSE BABY

Around the same time as the fatty liver diagnosis, I began to notice a small wobble just above my belt line, a slight extension to the meat verandah. When I slapped my stomach, it rippled like a wave. I imagined an ant body-surfing me. (The Waistline Masters?) This was new to me, and it brought up a whole range of vanity issues. I was constantly aware of my little fructose baby; it even affected my gait. Such is Zoe's magnificently positive outlook though, she reassured me by saying: 'I love it – now there is more of you to cuddle.' I wondered if she would still think that way after another five weeks. Would she even be able to get her arms around me for a cuddle?

What was happening to my body suddenly became quite serious and it took the idea of the film and the adventure to a whole new level. It's important to remember that both the fattening of my liver and the creation of the terrific 'ant surfing comp' on my stomach was done without any soft drink, chocolate, confectionery or ice-cream. If these supposedly 'healthy' foods were having this effect on me, then what impact were they having on other sections of the population? I packed a bag, kissed Zoe and her pregnant belly, cashed in some frequent flyer points and then had a nap. I woke up, had some sugar and then took off in search of some bittersweet truths.

In 2007, the population of just 360 people in Amata consumed 40,000 litres of soft drink.

SUGAR AND ABORIGINAL AUSTRALIANS

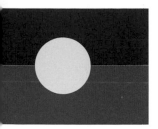

The image of mothers feeding their babies Coke had stayed with me long after my visit to David's hometown. I wanted to return to an Aboriginal community during my experiment to find out if the situation had improved. So eleven years later, I found myself in Amata, a small town on the APY (Anangu Pitjantjatjara Yankunytjatjara) lands with a cameraman in tow.

In 2008, Coca-Cola revealed that the Northern Territory was their highest sales region per capita in the world. This high figure was due in large part to demand from the Aboriginal stores.

Amata is located in the northwest of South Australia, but it lies a solid throw of a full Coke can from the Northern Territory border and the magnificent Uluru. In 2007, the town had a population of just 360, 92 per cent of whom were Aboriginal people. Those 360 people consumed 40,000 litres of soft drink in that one year.

It's not like anyone can just wander into these towns and start filming, or even just look around. Great respect must be shown to the local people and permits are required. We were lucky to gain access due to the generous efforts of John Tregenza, an Aboriginal man who has worked in Aboriginal health since the early 1970s. John is one of those rare beings who glows with compassion; you suspect he may well have been sent to this planet by some galactic peacekeeping organisation. He estimates he has driven more than 1 million kilometres while working for the health of his people. His nickname is 'The Chainsaw'. Why? Because he cuts through the bullshit.

OVERPRICED STORES

In 1998, John, along with Steph Rainow, their colleagues, and a local health council, set up a study into the nutrition of people on the APY lands. John told me that when he came to the area in the 1970s, just 10 per cent of food was bought from the local stores, compared with almost 100 per cent today. This is due to many environmental and political factors, which have affected the local people's ability to hunt or search for bush tucker (the main one being the introduction of a foreign grass that took over the plains and drove away the native plants and animals).

In the late 1990s, the local stores were owned by the Aboriginal people but were unregulated and many managers, often white people, took advantage and began to charge very high prices. John and his team reported that the locals simply could not afford to eat well; two days a week they didn't eat at all.

John, Steph and their team decided an overhaul of the stores was essential: the managers needed to be monitored and regulated, with health taking priority over profit. Most importantly, people needed to be educated about food and nutrition. The Mai Wiru ('Good food') policy was implemented, the first nutrition plan of its kind in the area.

The Mai Wiru team began to make radical changes to the seven stores in their care. They removed the deep-fryers and provided free cold water fountains, plus they made sensational rock music videos for the kids, featuring psychedelic vegetables and catchy lyrics about the perils of junk food. They brought in nutritionists who demonstrated the negative health effects of soft drinks by counting out the teaspoons of sugar contained in each bottle. The locals started to understand the effect that sugar was having on their health; they held a meeting and voted to have full-strength Coca-Cola removed from the town. The people had decided and, as John said, this was more important than anything. After years of being told what to do by governments, this action was coming from the ground up and it was empowering. By 2008, the Mai Wiru policy was in full swing and effecting real change. Amata had the lowest sugar consumption in the area.

EMPOWERMENT

Then, in 2009, the government became involved. They looked at Mai Wiru and a similar project, the Arnhem Land Progress Aboriginal Corporation (ALPA), and decided they would set up their own model, the Outback Stores. The Mai Wiru would become part of the government initiative and join the Outback Stores. The local people were hugely disappointed by this development: Mai Wiru was *their* project, it was working and it empowered them. Subsequently they refused to join Outback Stores.

The Mai Wiru's funding was then significantly slashed. While the government invested $90 million in the Outback Stores project, the Mai Wiru received just $150,000. John told me the first person to go was the nutritionist, the one person working directly with the locals to help them understand the relationship between food and health; someone who was making a real difference on the ground. As John explained: 'What we really need are nutritionists to come up and train some of the younger people. Those with the education can then stand in the stores and assist others as they are shopping, helping them to spend their money in the most beneficial ways. You cannot do that from Canberra.'

As a result, the sugar education stopped at Coke and many people in Amata now consume vast amounts of sugar in other forms. There may have been no Coke in the stores on our visit, but in its place were cans of Sprite and Lift, fruit cordials and juices, plus real sugar by the bucket load (literally). Something was still not getting through.

A recent study showed that 25–30 per cent of daily energy intake in some Aboriginal communities came from sugars.

'OTHER STORES I HAVE WORKED IN, THE SUGAR CONSUMPTION IS ASTRONOMICAL, THE PROBLEM IS KEEPING UP WITH THE DEMAND FOR COKE, YOU CAN'T STOCK THE FRIDGES FAST ENOUGH. BUT HERE AT AMATA, IT IS DIFFERENT. THE ELDERS GOT TOGETHER AND SAID, "WE DON'T WANT COKE", SO THEY DON'T HAVE COKE. I THINK THE MAI WIRU ORGANISATION COULD TEACH A LOT OF OTHER COMMUNITIES.'

MORRIS, AMATA STORE MANAGER

**JUST ONE CAN OF SOFT DRINK
A DAY INCREASES YOUR CHANCES OF
DEVELOPING DIABETES BY 22 PER CENT.**

EUROPEAN DIABETES JOURNAL

While Amata still consumes less sugar than other communities in the area, a report from the local nutritionist revealed that some Aboriginals were consuming up to 66 teaspoons of sugar a day. Is it any wonder that obesity, type 2 diabetes and renal failure are at chronic proportions within the Aboriginal communities? (Rates of type 2 diabetes are three times higher in Aboriginal people than other Australians and their rate of kidney failure is ten times higher.)

It must also be noted that the town of Amata has always been an alcohol-free zone. As John told me: 'There is a perception that these health problems are all alcohol related but they are not. Here in Amata they are directly related to poor diet.'

When John and local people asked the government for subsidies for fresh fruit and vegetables, the government refused. (In 2013, a report from the Australia Institute showed that the mining industry received $4 billion in government subsidies.) 'People don't laugh anymore,' said John. 'We had hope when the land rights came in but that has faded. I really worry about the future of my people, the old ones are dying and the young ones don't have anyone to learn from, all their learning comes from the television.'

John told me he is sick of attending funerals. He says the dialysis centres in Alice Springs are overflowing so people are sent to Adelaide. Dialysis involves four hours a day, three days a week, of sitting in a chair while blood is removed, cleaned and put back in. The costs are enormous, financially, physically and emotionally. John said many people choose not to have treatment; they would rather stay with their families and deal with the consequences.

I interviewed many of the community elders and all of them talked of a time when the only sweet foods they ate came from the sack of a honey ant or from the nectar of a flower stirred in water. Today they can open a fridge and get ten times the amount of sweetness in one cold can.

'Nobody told us,' said Leonard Burton, who is now on dialysis in Alice Springs Hospital. 'Sugar bad. Sugar no good. Before we had bush tucker, then the white man came and he brought lollies and biscuits. We all got sick.'

The Mai Wiru policy proved that with the right level of education and empowerment of the people, real change can occur. As John put it, 'It is a change far greater than a white man in an air-conditioned car visiting the town every few weeks and ticking off a checklist.' (Please see page 231 to find out how you can help the Mai Wiru.)

THE USA: HOME OF THE BRAVE, LAND OF THE FRUCTOSE

After one month and 1,240 teaspoons of sugar (the images of the Amata community still firmly in my mind), I decided I needed to speak to some people who knew a lot more about sugar than I did. It was time to take a trip to the sugar and processed-food capital of the world, to the centre of cultural convenience and the reigning number one world obesity champion – the USA.

After a fourteen-hour flight on which I was served a 'Savoury' snack box that had the word 'savoury' written in Nutella, I stopped by the local juice store to get a taste of what the USA had in store for me. I ordered the 'Strawberry Surf Rider', which came in a brightly coloured cup and looked incredibly healthy. It was packed full of strawberries, peaches and limes and topped with lemonade. It also contained 139 grams or 34 teaspoons of sugar. Now to consume that much sugar from whole fruit you would have to eat four peaches, 30 strawberries, 30 lemons and 9 limes. Of course, the fibre contained in the fruit would make this incredibly difficult to do. Strawberry Surf Rider may be fine if you are a bear about to hibernate for the winter but it can have deleterious effects if you are trying to make a 'healthy' choice for you or your children.

My adventure in the USA began with a road trip through the states of Alabama, Tennessee, North Carolina and Kentucky. The people of rural USA are quite up against it when it comes to nutrition. Prescription pills must be sold like candies here. As I clocked up the miles on the scenic freeways, I spotted some kind of fast-food outlet every 15 minutes – and none of them were 'The Broccoli Barn'. At first, I found it incredibly difficult to reach my 40 teaspoons of sugar a day by keeping to 'healthy' foods and avoiding fat. However, I was saved by a chain of chemists that sold enormous buckets of low-fat yoghurt, cartons of orange juice made with '5 per cent fruit juice' and packets of dried apple pieces emblazoned with the slogan: 'Soaked in fructose, nature's sugar!'

COKE IS NOW THE SECOND MOST RECOGNISABLE WORD IN THE WORLD ACROSS ALL LANGUAGES. IT IS SECOND ONLY TO OKAY.

STRAWBERRY SURF RIDER
contains 139 grams or
34 teaspoons of sugar.
Now to consume that much
sugar from whole fruit you
would have to eat
4 peaches,
30 strawberries,
30 lemons and
9 limes.

MOUNTAIN DEW MOUTH

Edwin Smith, aka Saint Denture.

At one point, it all became a bit too much. After one particular visit to a yoghurt café (with flavours such as 'birthday cake' and 'salted caramel pretzel') I slumped onto the grass in a parking bay and threw up a flavour of yoghurt that will never appear on the café's menu. This moment of poise and class was compounded when I looked up and saw a very respectable family of four watching me through the window of the café. I had a quick nap on the grass, pulled myself together, stomached some more sugar and hit the highway.

Through my sugar and jetlag haze, the countryside was even more spectacular and surreal than usual. An incredibly coloured 'Fanta' sunset poured out before me as I pulled into the small town of Barbourville, Kentucky, not far from the Appalachian Mountains.

SAINT DENTURE

I was in Barbourville to meet Edwin Smith. Edwin may actually be a saint and if he was ever to be officially recognised by the church, he should be known as Saint Denture, for Edwin is the master dentist in the area and he is a very, very busy man.

First, some background. In the great 'Cola Wars' of the 1990s, Pepsi and Coca-Cola were engaged in a fierce battle. Many lives were lost (and many are still being lost) but the Pepsi army took a strong foothold in the state of Kentucky and their number one sniper is the drink Mountain Dew. Now, despite its innocent-sounding name, in a 1.25 litre bottle of 'The Dew' you will find 37 teaspoons of sugar and 40 per cent more caffeine than the equivalent-sized bottle of Coke. Pepsi argues that in moderation, Mountain Dew is part of a healthy, balanced diet. I would argue that any drink containing that much sugar and caffeine is actually quite difficult to consume 'in moderation', nor is it designed to be.

In Barbourville, the locals have a habit of pouring 'The Dew' into their babies' bottles and feeding it to them. Many children under the age of twelve drink several cans a day. As Edwin told me, 'Their teeth just bathe in it all day long and so the enamel gets worn away very quickly.' Unfortunately, the kids' teeth quickly become black rotting stumps – and so Edwin and his colleagues termed the phrase 'Mountain Dew Mouth'.

Alarmed by the rising number of 'Mountain Dew Mouth' cases – and aware that many of those who needed help simply weren't able to access it – Edwin converted two motor homes into mobile dentist clinics. He now travels the state of Kentucky, fixing children's teeth and repairing the damage caused by soft drinks and juices. It is shocking to discover that the majority of Edwin's clients are under sixteen, with a huge number around the age of just three to four.

I spent a day with Edwin and his wonderful family (his young daughter was obsessed by my funny accent; little did she know I was just as transfixed by hearing a six-year-old with a Southern drawl). We picked up seventeen-year-old Larry, who has drunk Mountain Dew his whole life and was still drinking 'about twelve cans a day'. Larry was due to have 26 teeth extracted by Edwin who told me, 'I would love to say this is rare, but it isn't.' Now I have seen some gruesome horror movies over the years but all of them became *The Notebook* in comparison to what unfolded in the back of the motor home.

In the end, Larry could not go through with the full extraction. The poor fella just found it all too much and the anaesthetic wouldn't successfully penetrate his infected gums. (He had the op in the end though: Edwin wrote to me recently, enclosing a photo of Larry with his brand-new beaming smile. Mine was just as big when I saw the photo.)

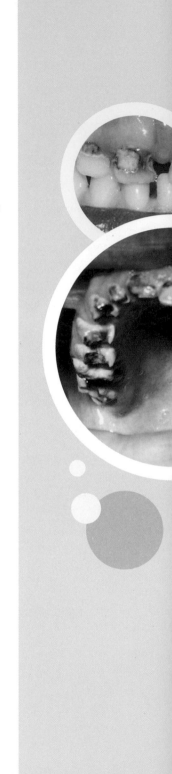

ENEMY NUMBER ONE

I think everybody accepts that sugar is the number one enemy of teeth. (Put simply, bacteria in our mouth called *Streptococcus mutans* loves the combination of fructose and glucose – when we consume sugary drinks or foods, it just fuels itself up and goes about its destructive business.) The trouble is that not everyone realises sugar is not just found in soft drinks, but is also in juices, sports drinks and 'organic' flavoured water. When Larry's mum told me she now realises she should have been giving her son blackcurrant juice as a baby instead, I had a flashback to my own youth and nearly short-circuited in front of her.

As I climbed into the car and imagined the vigorous brushing I would give my teeth that night, Edwin said to me, 'Damon, I see what sugar does to people's enamel and enamel is one of the hardest parts of the body. I can't imagine what it's doing on the inside.'

But despite sitting through one of the most painful and graphic scenes I am likely to witness, the biggest shock came when I asked Larry if his Mountain Dew days were done. 'No way,' he replied. 'I love it. I'll probably have a can when I get home.'

'I SEE WHAT SUGAR DOES TO PEOPLE'S ENAMEL – ONE OF THE HARDEST PARTS OF THE BODY – I CAN'T IMAGINE WHAT IT'S DOING ON THE INSIDE.'

The locals have a habit of putting 'The Dew' into their babies' bottles and feeding it to them.

LARRY

Many children under the age of six in Kentucky consume several cans of Mountain Dew a day.

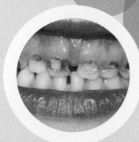

THE BLISS POINT

After seeing the effects of sugar on an Aboriginal community and then hearing Larry say he would keep drinking Mountain Dew despite the car crash in his mouth, I knew sugar had to have more of a hold on us than perhaps we realise.

I arranged to travel to New York to meet Michael Moss, the Pulitzer Prize–winning author of *Salt, Sugar, Fat: How the Food Giants Hooked Us*, to learn more.

And so it was that on a baking-hot day on the twenty-fifth floor of the *New York Times* building, my tiny suburban mind was blown for the second time that trip. 'Damon, you cannot underestimate the amount of science that goes into the design of food products,' Michael said. 'Nestlé has 700 PhD scientists trying to maximise the allure of their products; understanding the fundamentals of why we crave, how we taste and what makes foods addictive.'

Moss then shared with me some of the food industry lingo he had picked up when writing the book, terms companies use behind closed doors. Does a product have 'moreishness'? Does it have 'craveability'? How is its 'mouthfeel'? My personal favourite was 'vanishing caloric density'. This term applies to foods such as Cheetos or 'puffs', which feel like they disappear when you bite into them (the industry loves 'vanishing caloric density' because it allows the consumer to keep stuffing food into their gobs with the illusion that they aren't eating much) . However, my already pixie-like ears pricked up even further when Michael Moss told me about the 'bliss point', a term invented by a legend of the processed food industry, Howard Moskowitz. This is the man the biggest companies in the world turn to when sales are flagging. I had to meet him.

Howard Moskowitz's personality was as colourful as the wallpaper in his home – I have never seen wallpaper quite like it. During our interview, he was reluctant to mention any names or products, but in-between takes he would burst into song or rhyming couplets about his time in the industry.

THE IDEAL AMOUNT OF SUGAR

In the 1970s Howard discovered that each of us has a 'bliss point' for sugar. Simply described, this is our individual tolerance for sweetness or our ideal amount of sugar: the tipping point before something becomes too sweet. For example, you might enjoy one sugar in your coffee and like two even more. However, if you add a third sugar, the coffee becomes too sweet to drink. So your optimal amount of sugar – your bliss point – is two sugars. Howard went on to apply his bliss point principle to soft drinks; in

Moskowitz discovered that if you kept adding sugar to a product, people liked it more and more – but then they peaked and if any more sugar was added, they stopped liking it.

particular, he helped to make cherry-vanilla Dr Pepper a huge hit. His bliss point theory then marched its way around the supermarket and is now applied to an array of items including pasta sauces, bread, condiments and low-fat yoghurts. However, I got the feeling that these days Howard has ambivalent feelings about his discovery and its place in the food industry. As he wandered to the bathroom during a break in our interview, I heard him rap: 'The mass will not know, but ingest it quickly, loaded with sugar, it will make all of them sickly.'

I decided it was time to discover my own bliss point. I headed for Philadelphia and the Monell Chemical Senses Center, a cutting-edge research facility that boasts some of the finest brains on the planet. 'A clubhouse for PhDs,' says Michael Moss.

As I entered the foyer, I was immediately struck by the plaques on the wall: Kraft, Pepsi, Coca-Cola and a host of other food companies all make financial contributions to the centre.

I was told off twice within three minutes of entering the building. First, for asking about the nature of Coca-Cola's financial contribution and, secondly, for mentioning the very term I was here to investigate, bliss point. 'You must be very careful with language, Damon,' said the media officer. 'In here we call it the "sucrose preference test".'

THE TEST

The sucrose preference 'bliss point' test was very simple. It involved drinking cups of water containing various amounts of added sugar that I couldn't see and then choosing the one I liked the most. My results revealed I had the sweet tooth of a ten-year-old – perhaps no surprise given I was eating 40 teaspoons of sugar a day!

Interestingly, just a couple of months after removing sugar again, I found a banana too sweet. In fact, five months after stopping the experiment, I took the sucrose preference 'bliss point' test again and the results revealed my tolerance for sweetness had reduced considerably. I believe the palate adjusts after removing sugar, but the scientists at Monell told me that no such study has been performed. Perhaps an insight into how rarely people actually stop consuming sugar?

At Monell I also spent time with the scientists Julie Mennella and Danielle Reed, who told me humans have evolved with a love for sweetness. 'We are born to like sweet,' explained Julie. 'We may differ in the intensity we prefer, but it is our signal for energy.' I learnt that even newborns with undeveloped brains will smile when given sugar because the reward instincts are so primitive. 'When you put sugar in your mouth you can sense it all over your tongue, even at the back of the throat, even the sides of your tongue,' Danielle told me. 'Neural centres between the tongue and the brain tell the brain when you have put sugar in your mouth, like a telephone line.'

Perhaps this is why we are so vulnerable in the current environment. In the past, sweet foods were rare – wild fruit, honey and nectar – but they came packaged with vitamins and nutrients. Today our sweet hits are found on most street corners and are seriously lacking in nutrients.

The Monell scientists were lovely, but I felt the influence of the food industry at every turn. When I tried to spark a discussion about the effects of sugar on children, the media advisor shot back: 'What about video games? Should we ban video games for making children fat? Why is sugar any different?'

But I really knew it was time to leave when after two days of discussions (in which I had explained the experiment and made my thoughts on sugar very clear), the media advisor brought me a giant box of Dunkin' Donuts for afternoon tea. It was a gesture of goodwill, but flew in the face of every word that had come out of my mouth in the previous 48 hours. It made me realise how deeply entrenched sugar is within society and what a long road we have ahead when it comes to cutting back. Sugar is more than just an ingredient, we have it all wrapped up in rewards, emotions and celebrations.

Finding my bliss point.

SUGARBRAIN: OUR LOVE AFFAIR WITH THE SWEET STUFF

Early on in my sugar-eating marathon, I discovered how hard it was to feel full or satiated on the sugar diet. This only intensified the longer I stayed in the USA. I noticed I was now struggling to stay under the 40 teaspoons of sugar. I just wanted to eat more and more. But despite numerous enlightening conversations, I still hadn't found an adequate answer as to why Larry was still yearning for Mountain Dew despite his orthodontic horror show. What is it that makes us succumb time and again to foods or substances we know are detrimental to our health?

The current 'wisdom' surrounding people who overeat seems to be that they lack willpower or self-control. It's simple gluttony: they eat more than they need and they get fat. This belief has been pushed by the food industry for over 50 years and is not only bound up with judgment but is also devoid of any compassion or understanding.

In fact, recent research suggests that foods high in sugar can actually ramp up the brain's reward system to such an extent that it overrides an individual's ability to stop eating, while also creating subconscious patterns or eating habits. (Studies in rats have shown that reward circuits in the brain are so powerful that after eating high-sugar diets for three weeks, even when the rats are deliberately made sick on the same sugary foods, they still return to them the next day because the habit has been so strongly formed.)

To find out more, I headed to the Oregon Research Institute where a team of scientists has been using the latest technology to investigate the relationship between food and the brain. I was introduced to Eric Stice, one of the leading scientists at the institute. He told me: 'One of the interesting emergings in the literature says that the more you eat high sugary foods, the more you're going to desire sugary foods and want to keep eating them.' I think that was a very diplomatic way of saying that sugar may be addictive.

The scientists' plan for me involved taking high-resolution photos of my brain using a giant fMRI (Functional Magnetic Resonance Imaging) machine, while I drank a high-sugar low-fat chocolate milkshake through a large straw contraption. The team would take some happy snaps, then we would sit around and discuss them and perhaps even post some onto Instagram. #sugarbrain #milkshakehead #sweetbunsonaccumbens.

I put on my special glasses and was lowered into the jaws of the fMRI like I was a human sacrifice being offered to the 'Lord of the Machines'.

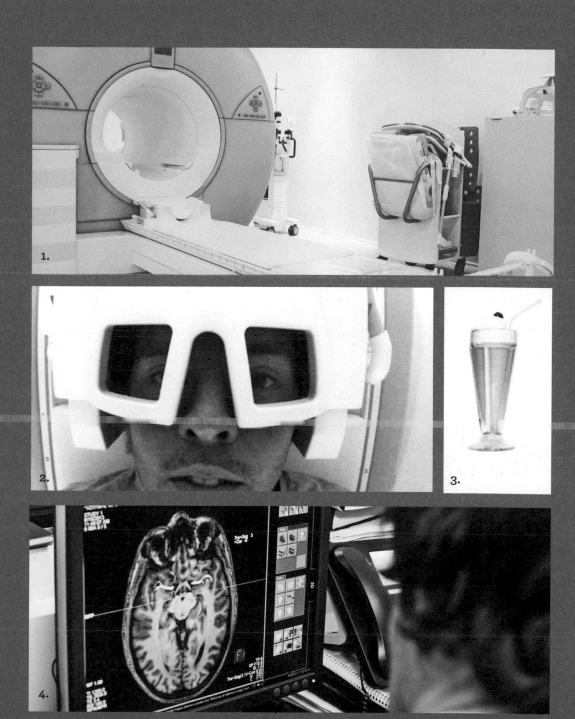

1. Lord of the Machines. 2. My special glasses that were clearly designed for giant-headed men. 3. The high-sugar low-fat culprit. 4. My brain saying 'cheese'.

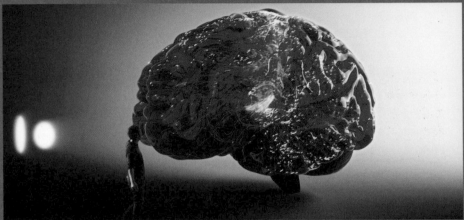

My brain lit up just by seeing a 'picture' of a milkshake.

Rats will work harder for sugar than cocaine. (I like to imagine a group of rats in a nightclub toilet doing lines of sugar!)

'DOES EATING PALATABLE HIGH-SUGAR FOODS ACTIVATE YOUR REWARD CIRCUITRY IN THE SAME WAY AS DRUGS OF ABUSE DO? THE ANSWER IS CLEARLY YES.'

ERIC STICE, OREGON RESEARCH INSTITUTE

For the next 30 minutes, I was shown images of a milkshake on a computer screen and then fed the milkshake through the straw. (I also viewed pictures of a glass of water and was given sips of water as a control.) I have to admit that being trapped in such a confined space and being fed warm sweet milk through a nipple-like straw contraption conjured up all sorts of emotions. The results of my scan were quite remarkable and showed that I was clearly enjoying my slightly twisted toddler flashback (see page 122).

The team in Oregon also told me about a study they had conducted on the influence of junk food adverts on children. In the experiment, children were monitored by the fMRI scanner while watching a TV show. The ads in the breaks were deliberately chosen and included skin products, mobile phones and junk food. All of the children's reward centres lit up when they saw the food commercials, with some children responding more than others. When the team conducted a follow-up a year later, they discovered the children with the higher reward responses had all put on weight.

'The more rapidly you learn those cues, the more likely you are to show future weight gain,' explained Eric Stice. 'Why do some people get sucked into a habit of overeating? Some individuals have hyper-responsive reward circuitry. It's more of a reward and they get sucked into the habit. Eric went on to tell me about a study from the University of Bordeaux that showed that rats will work harder for sugar than they will for cocaine. (I like to imagine a group of rats in a nightclub toilet doing lines of sugar!)

MODERN EATING HABITS

It is very easy to develop habits in the modern world. We live in an era of multi-tasking and many of us are juggling life rather than living it. How often do we put what we are eating at the top of the priorities list? How many of us have busily typed away at the computer while eating a sugary food without even thinking about it? We have got into the habit of mindless eating and this habit is taking its toll.

I can completely relate to this with chocolate and it is why I can't really have much of it anymore. If I eat it one night, I will want it again at the same time the following night. For me, a habit can form that quickly and before I know it, I'll be thinking about that sliver of dark deliciousness throughout the day, waiting for that evening's moment of pleasure. One thing leads to another and inevitably I'll rekindle my passionate love affair with the sticky date pudding.

It was the same when I was smoking. I knew it was doing me no good but my brain's reward system had overridden my sense of self-preservation and I kept needing the temporary high. This is how many people feel about sugar. These powerful urges have taken over and many of us are eating for reward rather than nutrition. People who are especially sensitive to sugar are the ones suffering the most. They aren't 'lazy'; they are just more vulnerable.

The good news, according to many experts, is that the brain can develop healthy eating habits just as quickly as it develops bad ones. I will discuss how I successfully changed my eating habits later in the book (see pages 153–155).

I recently had the great pleasure of spending a day with actor and encyclopaedic wordsmith, Stephen Fry. He has been a terrific supporter of the film and agreed to take part in both a sketch and also give us an interview. He spoke very candidly about his own battles with sugar addiction. For him, it began as a child when he would sneak down into the kitchen in the middle of the night to sprinkle sugar onto butter. He described these moments as 'almost being a sexual experience which produced glazed eyes and parted lips. It was an erotic moment for a child.' And for him this is what started the path of addiction.

He was very passionate about the topic of sugar and its effects on our health. He shared how he does have an addictive personality and does well for periods then can easily fall off the wagon.

He has particular concern with the abuse of power that the food companies display when it comes to labelling and food slogans and talked about how we will look back on this time in 30 years and be astounded by what we let them get away with.

What resonated most for me was his explanation of his own sugar addiction and how he copes. He said that everyone is different, his siblings don't have a sweet tooth and he could never understand this because he craved sugar so strongly. This has led him to believe that sugar addictions should not be seen as a weakness or a failure, but as a 'condition' of sorts.

'Sugar addiction is not a moral failing, it's a genetic condition, like asthma, which I also have. It's just the way I was put together, it's not my fault. On the other hand, I know what will give me an asthma attack and it is my fault if I bury my head in the trunk of a lime tree. And similarly it would be a bad thing if I constantly stuffed myself with sugary foods. Because I now know better.'

'I THINK IT HELPS TO KNOW THAT HUMANS CRAVE SWEET FOR BIOLOGICAL REASONS, IT'S NOT A MORAL FAILING. IT'S HOW WE DEAL WITH EXCESS.'

DANIELLE R. REED PHD — MEMBER, MONNELL CHEMICAL SENSES CENTER

'SUGAR ADDICTION IS NOT A MORAL FAILING,
IT'S A GENETIC CONDITION, LIKE ASTHMA...'
STEPHEN FRY, COMEDIAN, ACTOR, WRITER, PRESENTER AND ACTIVIST

THE 'A' WORD

Many of the experts I met on my trip to the USA talked about how
nervous food companies are of the 'A' word, addiction. The consensus
was that if sugar is proven to be addictive, it could unleash a legal
nightmare for the companies. There are scientific terms and principles
for what classifies a substance as being addictive, but the majority
of people I have talked to since starting this adventure care little for
the scientific wording. Perhaps those in the academic ivory towers
should pay a visit to a local 'overeaters anonymous' group. Sugar
gets a whole heap of airtime in those rooms.

'THE EASIEST WAY TO SEE IF YOU'RE ADDICTED TO SUGAR IS TO JUST STOP IT FOR 24 HOURS AND SEE HOW YOU FEEL. YOU'LL FEEL LIKE CRAP.'
DAVID GILLESPIE, AUTHOR OF *Sweet Poison*

'THERE IS NO WORD THE COMPANIES HATE MORE THAN THE "A" WORD – **ADDICTION**.'
MICHAEL MOSS, INVESTIGATIVE REPORTER AND AUTHOR

'SOME PEOPLE ARE VERY VULNERABLE, THOSE WITH A FAMILY HISTORY OF DEPRESSION, ALCOHOLISM, BUT EVEN PEOPLE WITHOUT THESE CONDITIONS. IF YOU GIVE THEM ENOUGH SUGAR YOU CAN CREATE THESE CONDITIONS AND GET THEM HOOKED.'
KATHLEEN DESMAISONS,
AUTHOR OF *The Sugar Addict's Total Recovery Program*

THE WORD SWEETHEART IS
DERIVED FROM THE 14TH-CENTURY
'SWETE HERT', MEANING 'FAST
BEATING HEART'. THE TERM BECAME
ASSOCIATED WITH LOVE AS IT WAS
LINKED TO SOMEONE WHO MADE
YOUR HEART THROB OR BEAT FASTER.

SWEETHEARTS

From Oregon I headed south to San Francisco. On the way someone suggested I visit the bakery that invented the 'cronut', a cross between a croissant and a doughnut. I was certainly keen to try one – until I found out people started queuing at 6 am, two hours before the bakery even opened. Yet again, I was amazed by the powerful hold sugar has on us. I'm pretty sure those people weren't lining up for the bread dough. For some of us, sugar is a real love affair.

My pursuit of understanding our relationship with sugar resulted in one of the highlights of my trip. In San Francisco, I met author Kathleen DesMaisons, who has been helping people to break their sugar habits for 30 years. Meeting her was like being in the presence of a wise elder; she brimmed with knowledge and shared it with humility and warmth.

Kathleen talked to me about the need for people to be kind to themselves when removing sugar from their diets, as for some people it can be a very difficult process.

She explained that sugar releases beta-endorphins in the brain, the very same endorphins that are released by love. The trouble is that while sugar creates a similar feeling to love, it is short-lived and a poor replacement for the real thing. Kathleen helps people not by telling them to quit sugar abruptly and instilling fear, but by suggesting they seek 'sweetness' in other areas of their life. She assists people in switching their 'rewards' away from sugar-based ones and replacing them with more nourishing and enriching rewards. When this is done successfully, she says, the need for sugar just falls away. She assures people they aren't losing anything by quitting sugar; in fact they are gaining a new way of life – one that should be relished and looked forward to.

When we understand that sugar releases the same endorphins as love, it becomes clear why chocolates are popular on Valentine's Day, why people can devour a whole packet of chocolate biscuits in a sad moment and how it can be very tough for some people to reduce their sugar intake. Break-ups are always difficult. Sugar is even trickier to split up with because it doesn't send you abusive drunk texts at 3 am – it just sits patiently in the cupboard in an alluring outfit, waiting for you to take it back whenever you're ready, no questions asked.

'People are turning to sugar for comfort,' Kathleen told me. 'We have forgotten how to get comfort in other places ... [When you] live without sugar, you start to see the wonderful things in life you didn't notice before, because you were too sugared up.'

Kathleen's words resonated deeply with me. So many things in my life got better when I removed sugar. The new 'sweetness' in my life came from a calmer energy, a brighter personality, a healthier appearance and a deeper connection with others. Because of this, I know refined sugar will never again be a regular part of my life.

Sugar releases the same opioids and endorphins that love does.
That's why breaking up with sugar can be very difficult for some people.

SUGAR AND SOCIETY

The longer I spent in the USA, the more I became aware of how sugar reflects much of our current society. Sugar is a quick hit, a cheap thrill devoid of any nutritional depth. These words could easily describe much of our modern culture – movie plots based on not much more than explosions and exposed flesh, instantly forgettable pop songs, E-news updates and flippant or 'quick to judge' comments on social media.

The Austrian philosopher Rudolf Steiner talked of a rise in materialism that coincided with the rise in sugar consumption. In other words, as we became obsessed with the quick fix of sugar, we also grew obsessed with the quick fixes of life.

Steiner's words make sense when you understand the rapid, fleeting metabolism of sugary foods compared with the languid metabolism of foods rich in fibre, healthy fats and nutrients. The latter is slow and calming, the former immediate and hyper. We now know that we humans are an energy field, in the form of a collection of rapidly vibrating atoms. Does it not make sense therefore that what food you choose to put into your 'energy field' will have an effect on it? Is it the gentle slow-release of an avocado or the quick burst of a chocolate bar?

ZOOKEEPERS IN DEVON, UK, HAD TO STOP FEEDING THEIR MONKEYS BANANAS FROM THE LOCAL SUPERMARKET. DUE TO GENETIC ENGINEERING THE BANANAS WERE TOO HIGH IN SUGAR WHICH WAS ADVERSELY AFFECTING THE MONKEY'S BEHAVIOUR!

A CALMER WORLD

What would society look like without many of us buzzing around, jacked up on sugar? Would we all slow down a little? Would we all have just a little more time for each other? Would we all be closer to understanding the true depths of what we are capable of individually and as a collective?

On my travels around the USA, I interviewed a retired NASA physicist named Thomas Campbell, a terrific chap who I could have happily talked to for hours (provided I had a box of muesli bars and some vitamin waters on hand). Thomas has a deep interest in the nature of reality and how we perceive the world. He helped develop a company called Monroe Laboratories that explores many facets of human consciousness. He decided to take a closer look at how sugar affects the brain.

Sugar can create a fog in our brain
that clouds our perception.

'WE NEED TO MOVE AWAY FROM A
SOCIETY THAT RESEMBLES A SHALLOW
GLOBAL SWAMP AND MOVE TO THE
DEEP LAKE THAT WE ARE ALL CAPABLE
OF BEING. THE REMOVAL OF SUGAR WILL
PLAY
A ROLE IN THIS PROCESS.'
THOMAS CAMPBELL,
RETIRED NASA PHYSICIST AND AUTHOR OF *MY BIG TOE*

An experienced meditator, Thomas once conducted an experiment where he got himself into a relaxed state and then consumed a variety of substances, including caffeine, nicotine, food colouring, sugar and artificial sweeteners. He told me that sugar had the most profound effect on him. He observed that the initial buzzing effect took around four hours to diminish and leave his mind and then a further eight days to leave his body. He now firmly believes sugar creates a fog in our brains that clouds our perception and our ability to operate at maximum potential.

I know from experience I am a different person when I'm not eating sugar. I am calmer, more patient, more present and more open to listen and communicate with people. Many would argue that they eat sugar all the time and they are fine, but how many of them have experienced what they are like without sugar? I suspect very few, given how early we begin our consumption and how prevalent sugar is in our food supply. Most people in our modern world would rarely go without sugar for four hours and even fewer people for the eight days it takes to completely leave our bodies.

This area of sugar and behaviour has huge implications for society and is only now being examined more closely. The main focus should be on children. With the huge amount of hidden sugar found in 'healthy' breakfast cereals, juices and muesli bars, is it any wonder that some children are struggling to sit still and concentrate in class?

'I HAVE HUNDREDS AND HUNDREDS OF PEOPLE GO OFF SUGAR JUST FOR THE REASON OF CLEARING UP THEIR MIND. I HAVE LOTS OF DATA ON THIS AND IT REALLY WORKS FOR PEOPLE. TYPICALLY THE RESPONSE I GET IS WOW! I HAD NO IDEA.'

THOMAS CAMPBELL,
RETIRED NASA PHYSICIST AND AUTHOR OF *MY BIG TOE*

'FOR THE FOOD INDUSTRY, THE HOLY TRINITY IS SALT, FAT AND **SUGAR**. BUT SUGAR IS THE BEST OF ALL BECAUSE IT IS A PRESERVATIVE, IT TAMPERS WITH THE APPETITE CONTROLS AND IT IS NOW SHOWING SIGNS OF ADDICTIVE QUALITIES THAT RIVAL SOME HARDER DRUGS.'

MICHAEL MOSS, AUTHOR OF *SALT, SUGAR, FAT*

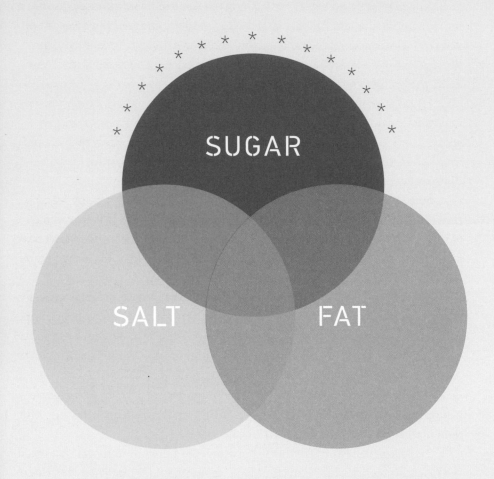

THE FOOD INDUSTRY: HOW DID IT GET SO SWEET?

It was now day 40, and after everything I'd learned during the experiment the main thing bothering me was: why wasn't the information about sugar public knowledge? And more specifically, why are guidelines about sugar intake so vague and nondescript? Being in San Francisco, I was in the right place to find some answers.

First, some background. On 24 September 1955, the US president Dwight D. Eisenhower suffered a heart attack and the issue of heart disease was thrust into the public domain. Debate soon began as to the cause of the disease and two strong camps emerged: one led by American scientist Ancel Keys, who declared fat was the problem; the other spearheaded by a British nutritionist, John Yudkin, who believed sugar was to blame.

Keys based his findings on studies from several countries linking fat consumption and heart disease. It was referred to as 'The Seven Countries' study. (It has now been revealed that the countries were carefully chosen and if he had included all 22 countries he examined, he would have had a very different result.)

John Yudkin, however, found that he could not make rats fat by feeding them fat, but he could make them fat by feeding them sugar. It is incredible to read his book *Pure, White and Deadly*, written at a time when he didn't have access to the scientific data or machinery that we have today. He was in many ways a prophetic figure.

The debate between the camps raged on for two decades until eventually the Keys' theory won out. Fat quickly became the villain and stole the headlines, while sugar was exonerated. By the late 1970s the low-fat movement was in full swing and seen as the healthiest diet – the food industry started tailoring their products accordingly. In San Francisco, the author Gary Taubes (*Good Calories, Bad Calories* and *Why We Get Fat*) explained to me that the iconic example is low-fat yoghurt: 'You remove the fat, add some fruit and sugar and then market it as a health product.'

Since that time and still today, the ingrained belief is: if you eat fat, you'll get fat and increase your chances of heart disease. John Yudkin and his sugar research fell to the wayside and Yudkin himself was ridiculed, so much so that other scientists were deterred from looking into the effects of sugar. If they did, they risked a similar fate.

MIRACLE INGREDIENT

With the food industry removing fat from their foods, sugar became a miracle ingredient. It was cheap, people loved it, it was a great preservative and importantly, it replaced the flavour and 'mouthfeel' that disappeared when the fat was removed.

In the USA this vilifying of fat also coincided with a team of Japanese scientists discovering that they could turn the glucose molecule in corn into fructose which has a much sweeter taste. This gave birth to the sweetener known as 'high fructose corn syrup'. Not only was this new ingredient cheap but it was embraced by a public who believed a sugar found in fruit couldn't be that bad. It never took hold in Australia but is still widely used in the USA, Hungary, Canada, Korea, Japan, Mexico and slightly less so in Germany, Greece, Finland and Portugal.

With sugar increasingly being used in our foods, it quickly became a very valued commodity. So much so that today it is worth $50 billion in global trade. And as often happens when these types of monetary figures are involved – the plot thickens.

In San Francisco, I met a lovely lady named Cristin Kearns (who sadly I couldn't fit into our documentary film). She is a former dentist who decided to delve a little deeper into the world of sugar and made a stunning discovery. In a Colorado library, she found internal memos and documents from a defunct sugar company that pertain to a key time in the history of the Sugar Association with links to the fat versus sugar debate.

In the mid-1970s while the debate was raging, the US government was reviewing the safety of sugar for the first time. Simultaneously, a committee investigating the role of sugar in the American diet was being put together by a Senator named George McGovern (later published as the *Dietary Goals for the United States*). McGovern's committee heard from various experts, including the scientist George Campbell who predicted an outbreak of type 2 diabetes if sugar consumption reached 70 pounds (about 32 kg) of sugar per person per year. (Americans now consume twice that amount and diabetes has risen 33 per cent in the past decade.)

With sugar firmly in the spotlight, the Sugar Association sprang into action. They hired the PR company Carl Byoir and Associates and then assembled a team of sympathetic scientists, led by a Harvard professor named Frederick Stare. (Stare was later outed for taking money from the food industry; he was also paid by the tobacco industry for a study denying the link between tobacco and heart disease.) Stare and his team released an 88-page document called 'Sugar in the Diet of Man', which they distributed to 25,000 carefully chosen organisations accompanied by a press release that read 'Scientists dispel sugar fears'. There was, of course, no mention that the 'research' had been funded by the Sugar Association.

In 1976, the Sugar Association won the Silver Anvil Award (the Oscars of the PR world) for 'excellence in forging public opinion'. They had seemingly convinced the public and policy makers that sugar was harmless. The great

From: THE SUGAR ASSOCIATION, INC.
1511 K Street, N.W.
Washington, D.C. 20005
Jack O'Connell/(202) 628-0189
or
Marian Rahl/(212) 986-6100

For Immediate Release

SCIENTISTS DISPEL SUGAR FEARS

WASHINGTON, D.C., August 4 ...
in the United States has reassu...
sugar, a pure carbohydrate, lik...
harmless when eaten in reasonab...

In 1974 these physicians an...
mounting importance as a food, ...
known about sugar's effects on ...
centers and universities across...
findings in "Sugar in the Diet ...
recently published in World Revi...
and available as a reprint.

SUGAR IS AN 'E...

Harvard's Dr. Fredrick J. S...
pounds of sugar consumed each ye...
about 80 pounds actually is eate...
changed in 50 years, contrary to...
is rising in the United States. ...

The Sugar Association, Inc.

1511 K Street, N. W. Washington, D. C. 20005

- - C O N F I D E N T I A L - -

July 17, 1975

TO: THE PUBLIC COMMUNICATIONS COMMITTEE

Gentlemen:

Shortly we will be mailing a packet to the press across the country containing: "Sugar in the Diet of Man," the Deutsch summary, a short news release and a cover letter signed by Bill Tatem.

We can anticipate some questions. Please be aware of the following answers, should they be put to you:

1. Who inspired the writing of the six papers?

   Ans. Dr. Fredrick Stare of Harvard suggested the need for this research and offered to organize their preparation.

2. Is Dr. Stare on your payroll?

   Ans. No.

3. Do you contribute to Harvard's School of Public Health?

   Ans. Yes, with an unrestricted grant. We also contribute to other organizations in the same manner.

4. Did you pay for the preparation of the papers?

   Ans. Yes. At the same time Dr. Stare asked us if we would be interested in the project, he asked us if we would be willing to fund it. We paid for research time, as we would with any research project, and purchased reprints.

5. Does the fact of your funding compromise the papers?

   Ans. No. We did not consider distribution of them until they had been accepted for publication by the journal, World Review of Nutrition and Dietetics. We feel the reputations of the doctors and their organizations and the prominence of the organ of publication attest to their accuracy and integrity.

Please hold and use for inquiries only.

Regards,

J. R. O'Connell
Director Public Relations

JRO:drb

Telephone: Area Code (202) 628-0189

The documents found by Cristin Kearns.

'THE SUGAR ASSOCIATION IS COMMITTED TO THE PROTECTION AND PROMOTION OF SUGAR CONSUMPTION. ANY DISPARAGEMENT OF SUGAR WILL BE MET WITH FORCEFUL, STRATEGIC COMMENTS AND THE SUPPORTING SCIENCE.'
SUGAR E NEWS 2003

'MANY OF THE REVIEWS WHERE THE FINAL REVIEW
SAYS THERE IS NO EVIDENCE THAT SUGAR IS
ASSOCIATED WITH METABOLIC DISEASE OF ANY
KIND ARE FUNDED BY THE SUGAR INDUSTRY.
IT'S REALLY SCARY TO EMPLOY STAFF AND KNOW
THAT MONEY IS GOING TO GO AWAY IF YOU FIND
AN ANSWER DIFFERENT FROM WHAT THEY [THE
INDUSTRY] WANT. I AM MUCH HAPPIER BEING A
STRUGGLING SCIENTIST, NOT KNOWING WHERE
THE MONEY IS COMING FROM . . . THAN A SCIENTIST
RECEIVING MONEY FROM AN INTEREST WHO IS
ONLY GOING TO BE HAPPY WITH ONE RESULT.'
KIMBER STANHOPE, UNIVERSITY OF CALIFORNIA, DAVIS

dagger in the heart for society was that the US Food and Drug Administration, the government body reviewing the safety of sugar for the first time, used this report as part of their findings. As the author Gary Taubes told me, 'They thought, how great! The science has been done for us.' Gary went on to say, 'The Sugar Association's president Jack Tatem stood up at a general meeting that year and said that there is no conclusive evidence linking sugar to these diseases and that this is the lifeblood of our organisation. If anyone links these diseases to sugar, then we're dead.'

Here is an excerpt from the Silver Anvil Awards stating why the Sugar Association won their PR award:

> Sugar, because of its universal usage and visibility, was a natural target for the lay nutritionists and promoters of fad foods and diets who appeared to capitalize on the concern generated by the consumer movement. As a result, the industry faced a barrage of criticism in the media suggesting that public consumption of ever-increasing amounts of sugar was responsible for a far-ranging variety of health problems. The objectives of the program were to reach target audiences with the scientific facts concerning sugar, enlist their aid in educating the consuming public, and to establish with the broadest possible audience the safety of sugar as a food.

(For more on this, see *Big Sugar's Sweet Little Lies* by Cristin Kearns and Gary Taubes, on the Mother Jones website: www.motherjones.com.)

Cristin informed me that this manipulation of evidence and deliberate muddying of the waters still goes on today. In 2003, when the World Health Organization (WHO) was looking to recommend that no more than 10 per cent of people's calorie intake should come from sugar, the Sugar Association openly attacked the organisation. In a statement, their president warned: 'We will exercise every avenue available to expose the dubious nature' of the report and urge 'congressional appropriators to challenge future funding' – in other words, they would lobby the US to pull US$260 million from WHO funding. Stern words, but when an industry is worth $50 billion, there is likely to be some bickering.

Many scientists I met talked about how publications funded by the sugar and beverage industries reported results that conflicted with their own research. In 2007, Kelly Brownell and his colleagues at Yale University published a study in the *American Journal of Public Health* that revealed the disparity in findings between an industry-funded study and an independent study in regards to the effects of soft drinks on the body. Another study in 2013 from Maira Bes-Rastrollo found that while almost all papers written without conflict of interest warned of the dangers of soft drinks, five out of six industry-funded studies concluded there was no danger.

Cristin summed it up well: 'I think it's important for people to understand industry tactics; some of the success of curbing smoking was when people started to understand what the industry was doing.'

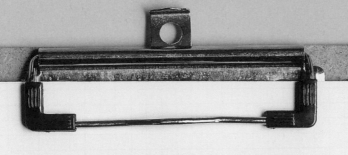

THE INDUSTRY PLAYBOOK:
IN CASE OF BAD PRESS AGAINST SUGAR, APPLY THE FOLLOWING TACTICS

Professor Kelly Brownell from Yale University has long been a crusader against the food and beverage industries and the dubious methods they employ to sway public opinion. He shared some of the industry tactics with me:

PAY SCIENTISTS TO FUND STUDIES THAT SUIT INDUSTRY POSITION

I experienced this first-hand. In Canada, I met a scientist who is funded by Coca-Cola (and other food companies). On my return to Australia, I saw an advert for a 'Sweet Drinks Symposium – a spotlight on fructose' – also funded by Coca-Cola. The same scientist I met appeared at the symposium to assert that there is no link between fructose and cardiometabolic diseases. Coca-Cola madly tweeted his comments; dietitians in the room re-tweeted them and then the host of the symposium told everyone to make a YouTube video about what they had learnt. It was an efficient machine to watch.

LABEL SCIENCE THAT DOESN'T SUIT THE INDUSTRY POSITION AS 'JUNK' SCIENCE

This pertains to what I mentioned earlier about studies done by industry versus the studies done independently and how they show glaring differences in results.

DISMISS CAMPAIGNERS AS 'FOOD FADDISTS'

Reject any movement against sugar as 'just another trend', and individual campaigners as attention-seekers or self-promoters simply trying to sell books. [Clearly this is what I am doing so if you are reading this then you have fallen for my plan, sucker!]

RUIN CREDIBILITY A classic industry move, where the integrity of the person who speaks out against sugar is questioned. [I may well cop some of this action so have fun trying to spot it!]

PAY MONEY TO PROFESSIONAL ORGANISATIONS SUCH AS HEART OR CANCER FOUNDATIONS
In Marion Nestle's *Food Politics*, she discusses how in the US, companies pay around US$4,500 to have a Heart Foundation tick appear on their product. In Australia, until recently, McDonald's paid the Heart Foundation AUD$300,000 a year to have a tick on items such as McNuggets and McChicken. Many children's cereals still have a tick, most of which contain high amounts of sugar.

RELEASE DISPARAGING COMMENTS IN THE PRESS ABOUT ANY MESSAGE AGAINST SUGAR

EMPHASISE THE 'CALORIES IN VERSUS CALORIES OUT' AND 'ALL CALORIES ARE EQUAL' DOGMA
This message is the backbone of the industry – if you get fat, it is your fault for being too lazy, i.e., it's your fault, not ours. However, recent studies show that calories behave very differently once they enter the body (see pages 117–118).

IT'S NOT ABOUT ONE FOOD!
This catchphrase goes: 'Let's not repeat the mistakes we made with fat and focus on a single item! We need to look at all foods!' A valid point, but given that sugar is now in 80 per cent of our foods, it practically is all foods!

IF ALL ELSE FAILS, PRETEND TO CARE
An example of this is Coca-Cola's 'Come Together' campaign, which focused on calorie consumption and exercise and how Coke wanted to help tackle obesity. Kelly Brownell believes this pretending to care comes out of a fear of litigation.

This is my reaction after consuming my 2,400th teaspoon of sugar. It was found in a child's school lunchbox (Apple Fruit Jelly) packet. Note the matching joy on Zoe's face.

THE LAST SUPPER

I was now 50 days into the experiment. Seven weeks of booster-rocket breakfasts, fructose tsunamis and ugly mood swings. I was physically exhausted and an emotional wreck. Cameraman Judd Overton and I had carried nine bags across eleven states of the USA in seventeen days. We had broken equipment, missed flights, been chased out of a Bible-belt town and heard the word 'sugar' more times than Willy Wonka. (I should mention that not one argument was had during the whole time, which given my sugar intake and mood fluctuations warrants miracle status. I doff my hat to you, Judd Overton.)

I missed Zoe and I also began to fear the baby would come early and the first impression our child would have of his or her father would be a twitching, flabby 'Shrek-like' grump with a fatty liver. It may well scream and scurry back in.

Now I'm sure Melbourne airport has seen some reunions in its time and the combined tears if collected would closely resemble Melbourne's rainfall figures, but the one I had with Zoe would be right up there on the airport staff's CCTV 'hugging highlights' tape. I will always remember the firmness of the hug and the bumping of two bellies that had both grown considerably in their time apart.

With just ten days to go before the end of the experiment, Zoe and I started to plan my 'Last Supper'. We agreed it should be something special, something poignant. We came across recipes for Jello Lettuce Salad and a Thanksgiving dish involving sweet potato and marshmallows, but in the end we decided that a major point of the exercise was to bring awareness to parents and their children about hidden sugars.

My time in the USA had made me painfully aware of how vulnerable children are to the effects of sugar – whether it be rotting teeth, brain fog, fatty liver or even the germs of addiction. By adding sugar to processed breads and cereals, juices, sauces and soups – not to mention all those 'healthy' lunchbox options – the food industry is teaching even the youngest children that most foods should taste sweet. This means that the healthy, savoury notes of foods like green veggies begin to taste 'peculiar'. The author Michael Moss described it as 'hijacking the biology of the child.'

1. My beautiful friend Gareth Davies actually made a pyramid out of the 2,400 teaspoons of sugar I had eaten. This was the second attempt after an ill-timed gust of wind. 2. It was far too easy to get to 40 teaspoons of sugar out of 'healthy' lunchbox snacks. 3. This was officially the last of my experimental teaspoons of sugar! It was disgusting.

I know from my own experience that a whole world of flavours opened up once I removed sugar from my diet. The constant sweetness of foods had a numbing effect on my taste buds, which meant that only the sweeter foods registered as enjoyable. When I did finally stop, it took a few weeks to enjoy Zoe's wholesome and nutritious cooking again.

Sugar-laden foods are robbing children of experiencing the natural, subtle flavours of healthier foods and are setting up taste preferences that may have detrimental effects later. We know that diseases of the metabolic syndrome (obesity, diabetes and heart disease) build slowly over time. The seeds of these diseases are being planted at a young age.

We decided my Last Supper had to demonstrate this point. I bought a standard school lunchbox and filled it with 40 teaspoons' worth of 'healthy' snacks, including Sesame Snaps, a jam sandwich, fruit drink boxes, sultanas, some kind of strange apple jelly snack, and yoghurt-covered raisins. I knocked them all off, high-fived the missus – then brushed my teeth for a whole ten minutes.

I was surprised at how emotional I was to finally finish my experiment. I could now let go. Although I knew the comedown in the days or weeks ahead would be tough, I looked forward to getting back to the person I was before. Sugar-free Damon was a lot happier and calmer. He was also much better equipped to be a father – and this brought huge relief to my family and especially to my beautiful, patient Zoe. The Sugar Daddy was no more.

The day after I finished, I headed straight to the needle house to meet with the clinical pathologist and have a vampiric amount of blood taken. Within those vials lay the answers to some sugary questions I had posed just 60 days before.

THE SCIENCE:
AND WHAT SUGAR
DID TO ME

PART TWO

THE SUGAR FAMILY ALBUM

Before we start, it is important to really get to know the members of the sugar family. For most people these sugars are like second cousins you only see at Christmas, but your health now depends on making an effort to get to know them. Some sugars are essential (like the aunty who always gives you fresh socks and undies), but others can have a damaging effect on our finely tuned bodies (for me it's the uncle with a crushing handshake and some inappropriate jokes about immigration).

SUGARS IN NATURE

THERE ARE FOUR MAIN SUGARS FOUND IN NATURE:

GLUCOSE is found in nearly every food we eat and is a key source of energy for our bodies. In fact, as far as our bodies are concerned, most things we eat are simply glucose in disguise. Bread, oats, rice, pasta, fruits and vegetables all break down to glucose in the body, so whenever someone talks about 'blood sugars', they are referring to glucose.

LACTOSE is a disaccharide, meaning it is made up of two sugars: glucose (see above) and galactose (galactose sounds like a giant planet made of milk where enormous cows are kings). Lactose is found only in milk and is often the first sugar we taste because it is contained in breast milk.

SUCROSE is also a disaccharide, being 50 per cent glucose and 50 per cent fructose. It is found naturally in sugar cane, sugar beet and most fruits. Importantly, it only makes up 10 per cent of a sugar cane but once refined, it becomes something that my nanna used to put a lot of in her tea – and the food industry loves to add it to processed foods.

FRUCTOSE is a monosaccharide, like glucose (meaning it is just one sugar). However, unlike glucose, which is found everywhere in nature and has long been part of our diet, fructose was once a rare sugar, found only in tree and vine fruits, some root vegetables and honey. These days, however, fructose is everywhere. As a result, while our bodies evolved to deal with glucose, we struggle to cope with the amount of fructose we now consume (see pages 109–110).

Our ancestors' fructose consumption included honey (from a beehive if you were brave), fruits in season, some vegetables, the sac of a honey ant. **Today's consumption includes fruit juices, soft drinks, sports drinks, ice-creams, chocolate, lollies, all foods containing refined sugar (roughly 80 per cent of processed foods), honey, fruit, dried fruit, sweeteners like agave and everything containing high fructose corn syrup.**

OTHER NATURAL SUGARS:

MALTOSE is made up of two glucose molecules and is found in germinating seeds such as barley. Maltose is found commonly in beer.

STEVIA is a natural plant native to Brazil and Paraguay. It is becoming increasingly popular as an alternative sweetener. Some people keep a stevia plant on their window sill and pull off a leaf when required, but the stevia you buy in a supermarket is often highly processed.

THE SUGAR FAMILY ALBUM

GLUCOSE

STEVIA

LACTOSE

SUCROSE

MALTOSE

FRUCTOSE

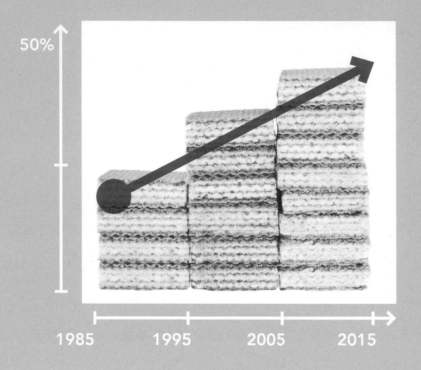

WORLDWIDE SUGAR CONSUMPTION HAS INCREASED 46 PER CENT IN THE PAST 30 YEARS.

CREDIT SUISSE REPORT 2013

When we eat certain foods, they adhere to a rule
They break down to GLUCOSE, which our body uses for fuel.
Breads, pastas, veggies and grain
All turn to GLUCOSE, to be used by our brain
But also our cells and our organs too
So without any GLUCOSE, there would be no you.

....

LACTOSE is next and she's as smooth as silk
She's the first sugar mammals have: it's in their mother's milk.
You'll find her in cheese and milk and tubs of yoghurt too
But take care if you're intolerant, coz you'll be rushing
   for the loo.

....

Now SUCROSE is the sugar that all the fuss is about
Some say it's dangerous, while others still have doubt.
SUCROSE is the table sugar we have in coffee or tea
She's the sugar family's sweet little daughter;
   or not, apparently.

....

And then there's FRUCTOSE – the sugar family's
   mischievous teenage son.
Because in the past, this FRUCTOSE was very very rare
It was found in fruits and veggies; and honey if you dared
But now we find it everywhere, in so many foods we eat
And you'll always know if you're eating FRUCTOSE:
   coz it's the thing that makes foods sweet.

PROCESSED SUGARS

SUCROSE is 'table sugar', the sugar I ate throughout my experiment. It is made up of 50 per cent glucose and 50 per cent fructose. Sucrose is now found in roughly 80 per cent of processed foods. It is made from either sugar cane or sugar beet. It comes in different forms, depending on whether the molasses is left in (brown sugar) or completely removed (white sugar) during the refining process. Despite brown sugar being perceived as the healthier option, once they enter our bodies all types of sucrose react virtually the same way.

HIGH FRUCTOSE CORN SYRUP (HFCS) is made by turning glucose molecules in corn into fructose. The fructose content of HFCS can range anywhere between 42 per cent and 90 per cent. Because it is cheap, it is used widely in sweetened drinks and foods, especially in the USA.

AGAVE is made from the bulb of the agave plant, found mainly in Mexico. It is often marketed as a healthy alternative, but given its extremely high level of fructose (90 per cent), it may actually be just as damaging as sucrose, or worse.

MAPLE SYRUP is made from the sap of maple trees and is 66 per cent sucrose.

COCONUT SUGAR is made from the flower buds of coconut palms. It is 70–79 per cent sucrose. It does retain some nutrients from the plant and studies have shown minerals, antioxidants and fibre present.

PALM SUGAR, often mistakenly confused with coconut sugar, is made from the sap of the palmyra or date palm.

RICE MALT SYRUP is made from the starch of brown rice. It is comprised of a glucose and maltose molecule and contains no fructose. However, it has a very high glycemic index because of the glucose molecules so will spike insulin quickly.

THE SWEETNESS SCALE

All of these sugars fall into a sweetness scale or ranking. Their different levels of sweetness are measured against sucrose (table sugar). Sucrose is given a sweetness rating of 1.0. It's the benchmark. (I have written in brackets how a ten-year-old might react to the taste.)

Sucrose 1.0 (More please!)

High Fructose Corn Syrup 1.0 (I said, More please!)

Fructose 1.4 (I really love you, Dad)

Glucose 0.7 (Boring)

Maltose 0.3 (Yuck)

Lactose 0.2 (Is this off?)

Stevia 300 (Whoah!)

When we eat sucrose, it is the sweet hit of the fructose molecule that really appeals to us. It is believed that fruits developed the sweet fructose in them to attract animals that could then eat the fruit and so release and spread the seed.

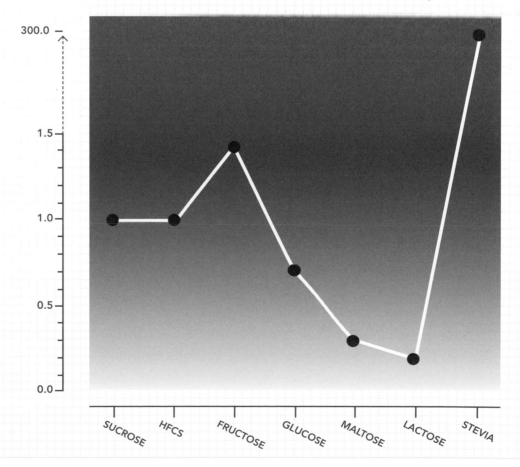

Queen Elizabeth I had such a love of sugar that her teeth rotted and blackened. *(This picture was made using caster sugar by the artist Marieka Walsh.)*

'SUGAR HAS BEEN ONE OF THE MASSIVE DEMOGRAPHIC FORCES IN WORLD HISTORY . . . BECAUSE OF IT, LITERALLY MILLIONS OF ENSLAVED AFRICANS REACHED THE NEW WORLD.'
SIDNEY MINTZ, AUTHOR OF *SWEETNESS AND POWER*

A HISTORY OF SUGAR

■ The first recorded knowledge of sugar comes from New Guinea in 8000 BC. A local myth of the time says the human race came about after the first man made love to the stalk of a sugar cane (an incredibly painful thought). In traditional New Guinean ceremonies, the priests sipped sugar water from coconut shells. Today they use Coca-Cola.

■ Sugar cane is taken to India and the Philippines in 6000 BC.

■ In 327 BC, sugar cane is discovered by Alexander the Great's general, Nearchus, who declares: 'A reed in India brings forth honey without the help of bees, from which an intoxicating drink is made, though the plant bears no fruit.'

■ In 500 AD, Indians begin manufacturing sugar by boiling the cane juice, which soon becomes known as the 'Indian luxury'.

■ From the 7th–12th centuries, the Arab world uses sugar for sweets, art displays and religious ceremonies. One report tells of a mosque being built from sugar and beggars being allowed to eat it when the festivities had concluded.

■ The Christian Crusades (beginning in 1095 AD) provide Europe's introduction to sugar. It becomes a rare and decadent commodity, purchased by kings. It is believed the contents of a single bag of sugar purchased at a supermarket today was all that could be found in the whole of 13th-century England.

■ In 1492, Christopher Columbus takes sugar cane to the New World. He plants the first crop in what is today known as the Dominican Republic. He becomes the first sugar grower to use enslaved Africans.

■ In the 16th century, Queen Elizabeth I loved sugar and had blackened teeth as a result. It was noted at the time that: 'The English poor ... looked healthier than the rich because they could not afford to indulge their penchant for sugar.'

- In the mid-1600s, the British settle in Barbados and Jamaica, which soon become known as the 'Sugar Islands'. In 1675, 400 English vessels travel to Britain with 150 tonnes of sugar on board. Between 1701 and 1810, 252,500 African slaves are brought to Barbados and 662,400 slaves to Jamaica to work on the plantations. It is believed that 40 million Africans were used as slaves in just 200 years, with the sugar industry the major customer.

- Early on, sugar is used as a spice. It then becomes common in medical practices, in the treatment of fever, dry coughs, pectoral ailments and chapped lips; it is also used in remedies to cover the foul tastes of snake skin or animal faeces. An English doctor is quoted in 1715 as saying: 'Sugar is a veritable cure-all, its only defect being that it could make ladies too fat.'

- Sugar really starts to dominate as a sweetener with the arrival of tea, coffee and chocolate. All of these are inherently bitter but add sugar and hello Starbucks. The first London coffee house opens in 1652 and by 1750 even the poorest English labourer's wife is having sugar in her tea. In 1760, *The Complete Confectioner* by Mrs Hannah Glasse is published, the first of its kind. Sugar is mixed with pastries and creams and becomes a dish to be had after savoury foods. What was a rarity in 1650 and a luxury in 1750 has become an essential by 1850.

- In the late 18th century, the First Fleet, en route from Great Britain to Australia, stops by South Africa, grabs some sugar cane and plants it in Port Macquarie. Today Australia is one of the largest exporters of sugar in the world. Charles Darwin on the *Beagle* in the 1830s tells of sailors giving Australian Aboriginal people sugar in exchange for dance performances.

- The 19th century is an incredible time for sugar. In 1800, 245,000 tonnes reaches European consumers. By 1890 the figure is 6 million tonnes. During this time sugar becomes known as 'white gold'. The French begin to use sugar beet for production, which does not require a tropical climate. Germany become the largest sugar beet manufacturer in 1880.

SUGAR WAS ONCE SO RARE AND PRECIOUS THAT AT PARTIES IN THE 18TH CENTURY, SOME PEOPLE WOULD CARRY SUGAR IN SMALL CONTAINERS AND WHEN THE NEED AROSE THEY WOULD TAKE A PINCH, LIKE A DRUG.

1. In 327 BC Alexander the Great's general Nearchus said, 'A reed in India brings forth honey without the help of bees.' 2. The sugar industry becomes a major contributor to the slave trade.

WHAT SUGAR DID TO ME

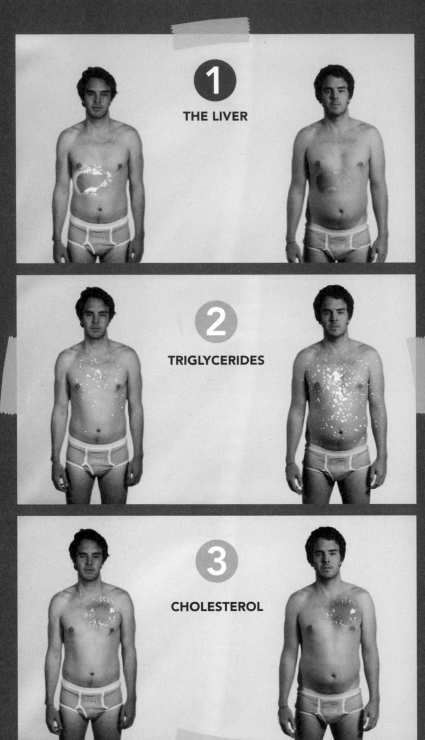

1 THE LIVER

2 TRIGLYCERIDES

3 CHOLESTEROL

1 THE LIVER: A key marker for determining liver health is an enzyme called ALT. Mine went from 20 at the start of the experiment (20 below the safety line) to 60 (20 above the safety line) in just two months. In Professor Blood's words: 'You went from the absolute best percentage of men to the worst in this short time.' (See page 109.)

2 TRIGLYCERIDES: This was a measurement taken to show how much fat I had in my bloodstream (triglycerides). In my case the liver was full of fat and was then pumping this fat out into my blood. This is how many scientists have linked sugar to metabolic disease. At the start of the experiment my reading was .08 and was very healthy but jumped to 1.5 after two months, which is regarded as the risk point (see page 115).

3 CHOLESTEROL: High levels of triglycerides are a new marker for potential heart disease. Cholesterol interacts with these triglycerides forming the dangerous small dense LDL particles that can clog and block our arteries (see page 124).

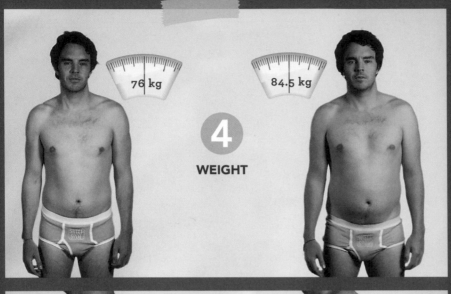

4

WEIGHT

76 kg   84.5 kg

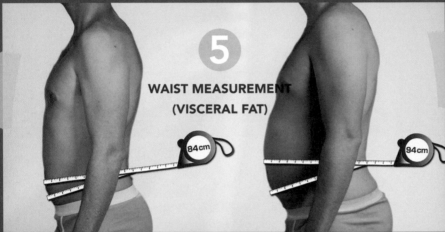

5

**WAIST MEASUREMENT
(VISCERAL FAT)**

84 cm   94 cm

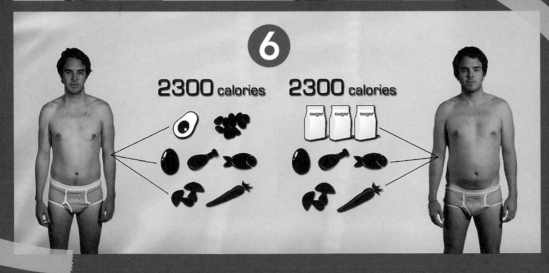

6

2300 calories   2300 calories

4 WEIGHT: During the two-month experiment I went from 76 kg to 84.5 kg. This was a total gain of 8.5 kilograms and an increase in my total body fat of 7 per cent. This happened without eating any junk food (see page 115).

5 WAIST MEASUREMENT: This was one of the more alarming results of my experiment. I put on 10 cm of fat around my waist. This fat is the dangerous type of fat called visceral fat that can cramp the organs and lead to many diseases of the metabolic syndrome.

6 CALORIE COUNT: If the waist measurement was alarming then this result was the biggest shock. I ended up eating the same amount of calories during the experiment as my previous diet. The big difference was the source of the calories. Pre-experiment I ate roughly 2,300 calories a day with 50 per cent from fat, 26 per cent from carbohydrates and 24 per cent from protein. During the experiment I also ate roughly 2,300 a day but 60 per cent came from carbohydrates, 18 per cent from fat and 22 per cent from protein. You can see that I virtually swapped healthy fats for sugar-laden products. New research suggests that the calories from sugar, and fructose in particular, behave very differently from other calories (see page 117 for a more detailed explanation).

UNDERSTANDING INSULIN

Before we find out the details of what sugar did to my body, we need to talk about insulin. For a non-scientist, insulin is the Rubik's Cube of hormones. I am not even going to pretend to understand all of its functions. I do know, however, that it controls what our body does with the food we eat, deciding whether to burn it for energy or to store it (this is known as 'fuel partitioning' in science speak).

Insulin and glucose (the sugar that most foods break down to) have a very close relationship. Insulin is vital for clearing glucose out of our bloodstream, either by allowing it to pass more easily into our muscle, organ or fat cells to use for fuel, or by using it to build glycogen, which can be thought of as a spare battery to call on when we need energy.

Insulin effectively acts as a key to our cells' door. It allows the door to open and the glucose to flow in to be used for energy. Insulin resistance is when the key gets stuck in the lock and needs a jiggle; the cell door won't open properly first time every time. This means more glucose is circulating in the bloodstream because it can't get through the door – and this can cause many problems, as those with type 2 diabetes know.

But insulin also plays a key role in regards to fat. It stimulates fat storage by keeping the fat in the cell.

This is where Gary Taubes' Carbohydrate Hypothesis comes into play. He, and many others, believe that when we eat carbohydrates like bread, pasta and sugar, our insulin levels increase in order to remove the glucose from our bloodstream. At the same time insulin is working to keep the fat in the cells, so our body never has the chance to burn off fat when insulin is dealing with all the glucose. Putting on weight is not only about too many calories and not enough exercise, but it is also due to a hormonal imbalance caused by excess carbohydrates.

Other factors can also raise insulin levels. When we are stressed, the hormone cortisol tells the body to prepare for fight or flight mode; the liver releases glucose, which in turn causes insulin levels to rise. Chewing gum stimulates saliva and tricks the body into thinking it's about to eat; insulin is then released in preparation.

(If you get the chance, watch a film online called *Cereal Killers*. A lovely chap named Donal cuts out all sugar and refined carbohydrates and so drops his insulin levels. He eats 5,000 calories (20,920 kJ) of healthy fats a day and still manages to lose weight.)

Insulin effectively acts as a key to our cells' door. It allows the door to open and the glucose to flow in to be used for energy. Insulin resistance is when the key gets stuck in the lock and needs a jiggle; the cell door won't open properly first time every time.

HOW SUGAR
AFFECTED MY MOOD

Right, so let's start at the beginning. What happened to me – and more specifically my mood – at the breakfast table on that first day of the experiment (the day I sprayed Zoe with vowel bullets from my mouth gun)?

The main fuel for the brain is glucose. As we now know, sucrose (table sugar) is made of 50 per cent glucose (with 50 per cent fructose).

That morning, as I ate my sugary breakfast of cereal, yoghurt and juice, insulin was very quickly released in order to deal with all the sugar (and refined carbohydrates) suddenly hitting my bloodstream. This is often called an 'insulin spike' (remember insulin is the key that allows glucose easier access into the cells in order to make energy). As the insulin rapidly removed the sugar from my bloodstream, this created a sugar crash (known as hypoglycaemia). My brain then wanted to get the blood sugars (glucose) back up because it needs them for energy and to function so it took action. It triggered the release of the hormone cortisol, which made me feel wide awake; this is what people describe as a 'sugar high' and is associated with 'fight or flight mode'. The brain also instructed the adrenal glands to release adrenaline (epinephrine) and noradrenaline (norepinephrine), which made me nervous and jittery, plus it engaged a neuro-transmitter called glutamate, which created the excitability and turned my brain into high gear. You can now appreciate why that seemingly healthy breakfast made me hyperactive and a bit unbearable to be around.

SUGAR AND MOOD

Throughout my experiment, I noticed that this 'up' feeling lasted for about 40 minutes before I would vague out, feel a bit numb, and then catch myself staring at nothing in particular. I would then open the fridge, pull out a perfectly named Sanitarium 'Up and Go' with 4 teaspoons of sugar and the system would be rebooted with chatty, energetic Damon making a return.

It is estimated that 50 per cent of the US population are hypoglycaemic in some form. Some people are more sensitive than others and when their blood sugar levels fall, they can feel anxious, angry, weak, grumpy, irritable or moody. They also tend to crave a sugary food for a quick fix to feel 'back on track' again, but really all they are doing is keeping the cycle alive. During my experiment, I understood very quickly that consuming high levels of sugar can have an enormous impact on mood and behaviour.

What's more, over time the adrenal glands engaged by the brain in response to sugar can become accustomed to producing more cortisol than is healthy. Particularly when combined with a stressful lifestyle (stress also stimulates the release of cortisol), a high-sugar diet can really wear out the adrenal glands and greatly affect their ability to function. Adrenal fatigue is becoming increasingly common and is associated with insomnia, tiredness and weight gain; some even speculate it may contribute to heart disease.

A study by Dr Lidy Pelsser of the ADHD Research Centre in the Netherlands was published in a Dutch journal in 2011. It showed many children could come off their ADHD medication by simply changing their diet. She discovered that 64 per cent of children in her study diagnosed with ADHD are actually experiencing a hypersensitivity to food.

A study by the University of Vermont in 2011 concluded that high-school children who consumed five or more soft drinks per week showed much higher levels of violent behaviour than those who didn't.

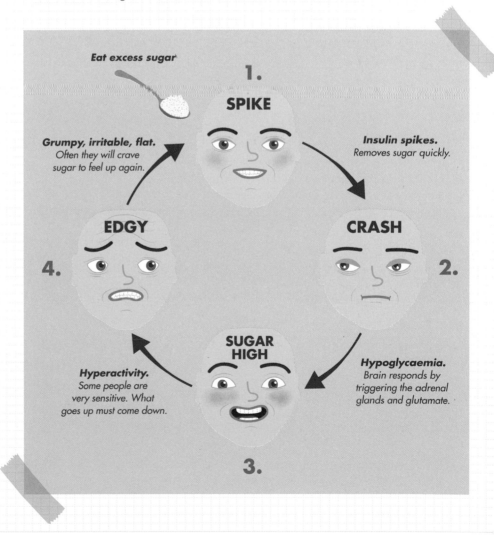

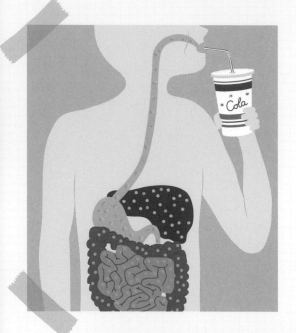

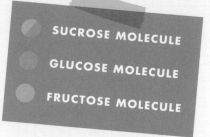

SUCROSE MOLECULE

GLUCOSE MOLECULE

FRUCTOSE MOLECULE

GLYCOGEN
GLUCOSE

GLYCOGEN
GLUCOSE

FAT

It's estimated 5.5 million Australians now have fatty liver disease. Only 6,000 of these are the result of alcohol. In the UK, the NHS says 25–30 per cent of people have early forms of non-alcoholic fatty liver disease.

HOW SUGAR GAVE ME
A FATTY LIVER

The moment of the experiment where I really knew that there would be a story to tell (and the moment that lit up the producer's face like a Coke sign in Las Vegas) was on Day 18 when I was told that my liver had turned to fat.

We know that sugar is sucrose, which is made up of 50 per cent glucose and 50 per cent fructose. When sugar enters our body, it heads straight to the small intestine. On arrival it is met by an enzyme that splits the sucrose into fructose and glucose. The fructose and glucose molecules are absorbed into the bloodstream and head towards the liver, which is an incredibly important organ for sugar metabolism and is a kind of 'sugar-sorting station'. On the way, the glucose is used for energy by any cell that needs it. The fructose, however, needs to be processed by the liver before it can be used for energy.

When it comes to glucose, the liver deals with it by either using it for energy or by storing the excess for when we are running a bit low. This form of energy storage is called glycogen and it is like our body's spare battery on a laptop. When we are connected and eating, we charge it up. When we disconnect, we first use up what's in the bloodstream, then we start depleting the glycogen stores (running on our spare battery).

FRUCTOSE TO FAT

When it comes to fructose however, the liver responds in a different way. There is an enzyme called fructokinase that is not regulated by liver energy status – meaning that it is actually 'switched on' all the time, so the liver pulls fructose out of the blood and hoovers it up regardless of whether it needs it or not. (It could be said that the liver has a massive crush on fructose and love is blind.) Scientists believe this is because throughout evolution, fructose was very rare in nature: only found in honey or seasonal fruits (remember; no processed foods or juice bars).

The fructose is also used for glycogen if needed, especially for topping up after a big sleep or after a marathon run . But once the battery is full, the fructose immediately gets turned into fat. This is what we do when we have too much energy – we store it as fat. Because the liver is switched on to fructose all the time, it just keeps pouring in and ramps up the fat. This is where the problems really begin.

This stuff can explode in your brain if you're not careful. It's important to understand this because fructose turning to fat in the liver can set off a series of chain reactions in the rest of the body.

Let's try to simplify this so we can understand better. Imagine the liver as a giant factory. This factory's prime objective is to sort sugars into energy or storage. One production line deals with glucose and on the other side of the warehouse, a production line deals with fructose. All along the conveyor belts are enzymes that help the process along (like workers in a car factory).

Now the glucose production line is extremely efficient. Workers here have been dealing with vast amounts of glucose over many years. 'Employee-of-the-month' awards line their walls. They have a terrific boss and the first thing he does is check and see if the factory (liver) needs to make any more energy. If more energy is required, the workers will tirelessly press on and get it done; if not, the boss pushes a button, an alarm rings and the production line shuts down. The glucose is then either stored out the back for later (as glycogen) or it heads to the muscles, brain or fat storage cells in the rest of the body (the dealerships).

Things are very different on the other side of the warehouse. The fructose production line has never had to deal with much of it in the past, so it has evolved without a boss to call the shots. As a result when fructose rushes in, there is no one in charge to check if the factory needs energy. The fructose just heads straight onto the production line. And when that line to make energy becomes full very quickly (because the glucose production line has also stored energy out the back), it opens up a new production line – a line that is really good at making fat. This fat then fills up the entire factory, having huge effects on all the production lines and eventually makes its way out of the factory doors and into the body metropolis (this is what happened to me – my factory filled up with fat after just 18 days).

Scientist Jean-Marc Schwarz, who I met in San Francisco, compares our current sugar consumption with the process of making foie gras, the practice of force-feeding ducks or geese large amounts of carbohydrates so their livers fill up with fat. The fatty liver is then enjoyed as a delicacy. Dr Schwarz says this is exactly what we are doing to our own human livers in our fructose-dominated world. Perhaps most disturbing of all is that we don't have to force-feed ourselves or our children because the sugar comes in shiny, cleverly marketed packages.

The liver factory.

The fizzy force-feeding foie gras farm for children.

HOW FRUCTOSE HAS AVOIDED THE SPOTLIGHT

There are a couple of reasons why we are only now starting to examine fructose with Hubble telescope-type detail.

The first reason fructose has kept a low profile is that fructose is from fruit – if it's from fruit then it can't be bad, right? This may be an accurate statement if you consume small amounts of it via some berries every now and again. However, when we flood the liver with fructose by drinking juice or soft drinks or by eating it in 80 per cent of our processed foods, things may start to look a little different.

The second reason is a little more complicated. Carbohydrates are often classed under their chemical structure as 'simple carbohydrates' or 'complex carbohydrates'. Simple carbohydrates are molecules of one or two sugars bound together and include fructose and glucose. Complex carbohydrates are chains of sugars that can be thousands of sugars long. As a result they take longer to break down. These include foods like vegetables, beans and lentils and are considered much better for you.

In the early 1980s, the Glycemic Index (GI) was developed to help us understand how quickly a carbohydrate is digested in the body and converted into blood sugar (which our cells use for energy). The GI ratings system acts like a food scoreboard. Foods that convert to blood sugar quickly (and spike insulin) are given a high score: a plain white baguette is 95 (out of 100), white rice is 89 and instant oatmeal is 83. People are encouraged to avoid these types of foods. (Remember that spiking insulin traps fat so these foods may actually be very fattening.)

Foods that take longer to convert to blood sugar and so in theory are kinder on our bodies get a lower score. They include the likes of chickpeas (10), apples (39) and ice-cream (36). Yes, ice-cream! It is important to know that the presence of fat or protein in a food decreases the GI score. This is because fat does not spike blood sugar levels so the score is kept lower. Based on this system, carbohydrates became very easy to define as good or bad and many health and nutrition bodies accepted the method.

The anomaly is that foods containing sugar can often have a relatively low score. Sucrose has a lower score than potatoes or white bread. Ice-cream gets 36, but green peas get 51. Are we supposed to think that ice-cream is the better option to have on our plate with steak and mash?

We now know that the reason for this is the fructose. Fructose gets a GI score of around 19, which is incredibly low. In fact, for some time it was highly recommended to diabetics. In the early 1980s, in the *New England Journal of Medicine*, a diabetologist, John Bantle reported that fructose could be the healthiest carbohydrate there is and wrote, 'We see no reason for diabetics to be denied foods containing sucrose (table sugar).'

If you have managed to get through the 'How Sugar Gave Me a Fatty Liver' section (pages 109–110) without falling asleep you will now know that very little fructose makes it into the bloodstream; instead it is hoovered up by the liver and turned into fat. So when it comes to table sugar or high fructose corn syrup, the GI score is kept low because the fructose isn't in the bloodstream to spike the levels, it's actually in the liver wreaking some other havoc! The fructose isn't in the blood, people; it's in the liver.

This is something we are only now beginning to understand, so you can see why Mr Fructose has been given an all-access pass to the World Sugar Party for the past 35 years. But the party's over mate – time for sleepy-sleeps.

The Glycemic Index scores foods out of 100. The score indicates how quickly the food is converted into blood sugar (and in turn how quickly it spikes insulin).

'IN SUGAR GROWING COUNTRIES THE [WORKERS] AND CATTLE EMPLOYED ON THE PLANTATIONS GROW REMARKABLY STOUT WHILE THE CANE IS BEING GATHERED AND THE SUGAR EXTRACTED. DURING THIS HARVEST THE SACCHARINE JUICES ARE FREELY CONSUMED; BUT WHEN THE SEASON IS OVER, THE SUPERABUNDANT ADIPOSE TISSUE (FAT) IS GRADUALLY LOST.'
THOMAS HAWKES TANNER, *The Practice of Medicine*, 1869

HOW SUGAR MADE ME FAT

There are two main types of fat in our body: subcutaneous fat, which lies just under the skin and is thought to be relatively harmless; and visceral fat, which lurks around the abdomen and organs and poses more of a health risk.

In just 60 days, I gained 10 centimetres of visceral fat around my belly and 8.5 kilograms in total weight.

Let's just remember this happened while I was on a low-fat diet consuming the same amount of calories as before I started and without any help from soft drinks, chocolate, confectionery or ice-cream.

So how the hell did I become so fat? Shouldn't I have lost weight? If I was consuming the same amount of calories, and not eating fat, how on earth did I put on weight?

There is still no definitive answer as to why fructose makes us fat, but this is the best explanation I was given on what happened to my abdominal region. The most important thing to remember is that fructose in the sugar was transformed into fat in the form of fatty acids and triglycerides, which entered my bloodstream. These triglycerides can also accumulate in adipose tissue (fat), which for me and most men, likes to gather just above the belt buckle.

This fattening of the liver can also lead to insulin resistance (see pages 104–105), which in turn leads to more sugar (glucose) in the bloodstream. The body then secretes more insulin to keep the blood sugar under control. In addition to this, the glucose half of the sugar is also circulating in the blood and also requires insulin to remove it. In the words of Jean-Marc Schwarz: 'It's a double bingo.' There is clearly a lot of insulin flying around. In fact, my insulin levels nearly doubled during the experiment.

Remember that insulin stimulates fat storage and keeps fat in the cells. If insulin levels are high, only the glucose in the blood is burnt for energy and the fat remains in the cells. We do not burn off the fat for energy. We store it. So in my body, the liver just kept making fat from the sugar and the high insulin levels it created told my body to hold onto that fat.

One scientist told me some lucky people can burn off fructose very easily, especially some teenagers. He said that others may appear quite skinny but if they are eating a high-sugar diet, the fat could be gathering around their organs. Because there is no external show of fat, they may be unaware of potential health problems. There is a term for these people: TOFI – thin on the outside, fat on the inside. If you eat a lot of sugar, it may be worth having your triglyceride levels checked (which is fat in the bloodstream) even if you're not overweight.

Kimber Stanhope of the University of California, Davis, recently conducted a study in which two separate groups were put on a glucose or fructose diet for ten weeks. Both groups put on the same amount of weight, but only those consuming fructose had increased fat in the liver, increased triglycerides and insulin sensitivity. The glucose group put on the safer, subcutaneous fat, whereas the fructose group put on the visceral kind, the fat that surrounds the liver and intestines and increases the risk of metabolic disease.

'THE BODY IS LIKE A BIG WEB. EVERYTHING IS CONNECTED AND WHEN YOU START TO SHAKE THE WEB ON ONE SIDE WITH SUGAR INTO FAT METABOLISM, THEN YOU START TO SEE SOME DETRIMENTAL EFFECTS ON THE OTHER SIDE OF THE WEB.'

JEAN-MARC SCHWARZ, TOURO UNIVERSITY, CALIFORNIA

WHEN A CALORIE IS NOT A CALORIE

According to the World Health Organization (and many others), being overweight is simply an energy imbalance between calories consumed and calories expended. The solution is apparently very clear: eat less and exercise more.

Now, I have always thought that the body is an intricate and complex system and although a calorie may be a calorie on a plate, once that calorie enters this intricate and complex system, it will behave differently according to all sorts of factors pertaining to the individual. Turns out this hunch may have some truth to it.

Research from the likes of Rob Dunn, a biologist at the North Carolina State University and Professor Robert Lustig, a paediatric endocrinologist and 'Godfather' of the anti-sugar movement, has shown that calories do a whole range of things once they enter the body, depending on factors such as how we prepare our food, the level of bacteria in our gut and even the different energy required to digest different foods. For example, proteins may take up as much as five times more energy to digest than fats because enzymes need the energy to unwrap the strings of amino acids that build the protein.

According to the label, a packet of almonds may contain 170 calories (710 kJ), but the body will only receive 129 of those calories (540 kJ). This is because the fibre in the almond delays absorption into the blood, plus the bacteria in our gut also chews them up.

Fats release 9 calories (38 kJ) per gram when burnt, but we now know that there is a range of fats that all behave differently. Omega 3 fats from sardines, salmon or walnuts are seen to be good for you, but 'trans fats' from potato chips, some margarines and doughnuts are very dangerous. However, you will just see 'calories from fat' on the label.

Kimber Stanhope's glucose versus fructose experiment (see opposite page) clearly demonstrates this point. Although the groups ate the same number of calories, the calories from fructose affected the body in a very different way from the calories from glucose.

What shocked me the most, and the others helping me throughout the experiment, was that I ate the same amount of calories on average during the 60 days as I did prior to the experiment. I ate quite a high-fat diet before I started, which included good fats like avocado and nuts. I thought my calorie count during the experiment would be much higher because of all the sugar, but I learnt that 1 gram of fat equals 9 calories (38 kJ); while 1 gram of sugar equals only 4 calories (17 kJ). As a result I can understand why we are told to avoid eating fat and that sugar may be okay – there's over double the calories in fat. It seems, however, that all the sugar I was eating was mainly turning into fat in my liver and as my insulin levels remained high, the fat was then being stored (as explained on page 115). Plus the fructose half of the sugar was affecting my appetite controls (see page 121). In my experience, the calories from sugars behaved very differently in my body from the calories from the foods I was eating before I began. Perhaps this could explain why some of my friends spend hours at the gym and count their calories, but still enjoy desserts and energy drinks and cannot seem to lose the weight.

ALL CALORIES ARE NOT EQUAL

My ears always prick up now when companies such as Coca-Cola push the message that all calories are equal and that weight-loss is all about exercise. This of course completely deflects any blame from the unique effects of sugar. Frederick Stare, the professor who was outed for taking money from the beverage industry and who led the Sugar Association's panel of experts (see pages 76–79) is quoted as saying, 'Calories are all alike, whether from beef or bourbon, sugar or starch or from cheese and crackers.' It does now make me wonder if it all may be disturbingly connected.

A good analogy by Sam Feltham, a great guy who has done many human lab-rat eating experiments on himself and runs a website called 'Smash The Fat', explained what may have happened to me. He compares the body to a kitchen sink. There is a tap for healthy foods and one for processed or fake foods like sugar and other refined carbohydrates, which I ate but normally don't. The calories from healthy foods, including certain fats come in and flow through easily as energy, maintaining balance in the body but the calories from sugar and refined carbohydrates affect our biochemistry by turning to fat in the liver and by also keeping insulin in our blood and when insulin is around we can't burn off fat. These fake foods are like limescale that can 'clog our sink'. Suddenly the calories coming in don't flow out as easily and instead get stored as fat which begins to accumulate where we don't want it and can lead to further damage.

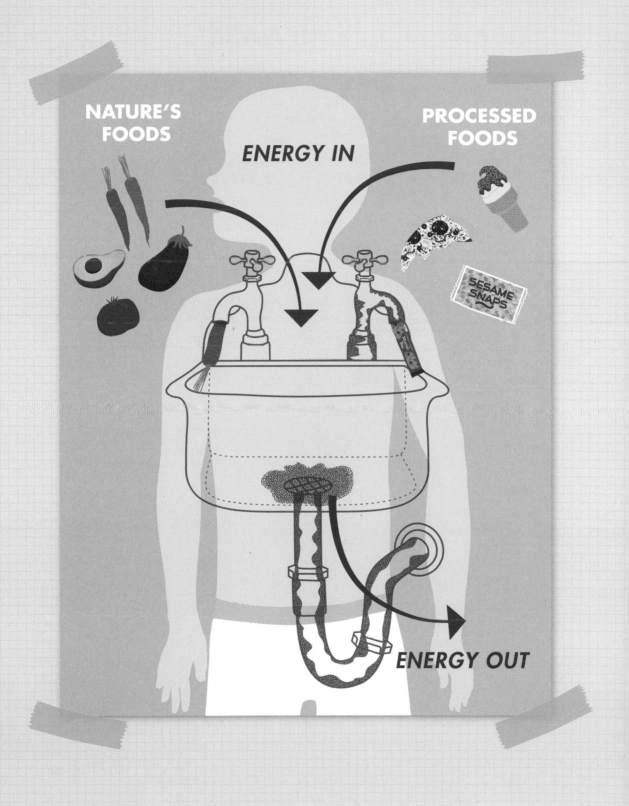

The hypothalamus is like the bouncer of a nightclub.
It seems fructose has the ability to sneak in unnoticed
like a cunning pimply teenager.

HOW SUGAR MESSED WITH MY APPETITE

To understand how sugar affected my appetite during the experiment, let's look at some research by Kathleen Page and her colleagues at the University of Southern California. The team looked at the effects of glucose and fructose on an area of the brain called the hypothalamus. The hypothalamus is our appetite control centre and responds to hormones (in particular leptin, which is sent from our fat cells) that tell the brain we are full.

Kathleen and her team reported that when their subjects consumed a drink containing only glucose, the blood flow and activity in the hypothalamus area decreased and the subjects reported feeling full. When those same subjects were fed a fructose drink, they did not report feeling full and their hypothalamus area remained active. In essence, the brain still thought the body was hungry.

I like to imagine the hypothalamus as the bouncer of a nightclub. He has a clicker that tells him when the club is full. It seems fructose has the ability to sneak in unnoticed like some cunning pimply teenager.

Endocrinologist Dr Robert Lustig believes that insulin may play a large role in blocking leptin, the hormone that tells the hypothalamus we are full. So even though we have plenty of sugar in our system and we are making fat in the liver, the brain still thinks we are hungry. Dr Lustig believes this is why it is so easy to have a glass of apple juice or soft drink with dinner and be able to keep eating. These are calories entering our body that our brain simply doesn't recognise.

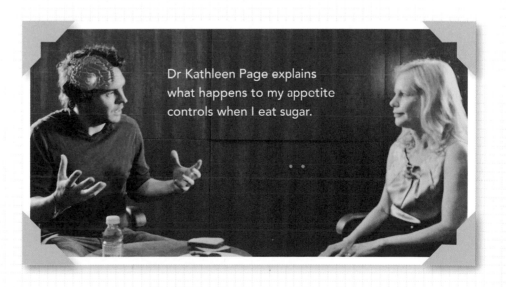

Dr Kathleen Page explains what happens to my appetite controls when I eat sugar.

HOW SUGAR AFFECTED MY BRAIN

So after spending 30 minutes in the fMRI machine in Oregon being fed a high-sugar low-fat milkshake and having toddler flashbacks, this is what the photographs of my brain revealed (see pages 60–63).

The first step was seeing the cue or trigger, which in my case was a milkshake. This can be the real thing or even a picture of a milkshake! When we see the sweet thing, a chemical called dopamine is released. This reaction evolved from the time when sweetness was rare – when we spotted something sweet the brain told us to get after it, as we needed it for energy.

Then we taste the sweetness (in my case, I drank the milkshake) and chemicals called opioids and beta-endorphins are released, which make us feel terrific. This is the sugar hit and our reward centres are flooded with good feelings. In fact, according to research, sugar lights up the same reward areas as nicotine, cocaine and sex. But it doesn't last long. Now if some of us eat sugary foods often enough and establish this feeling of reward, it can form a subconscious habit – one that is easily triggered by more images of sugary foods.

For me, the most shocking aspect of these results was how my brain responded when I simply saw the *picture* of the milkshake. My brain lit up nearly as much as when I actually drank the milkshake.

The consequences of this in regard to food advertising are obvious. We live in a world full of cues – everywhere we look we see billboards and TV ads featuring enticing images of sugary drinks and luscious ice-creams. Those individuals who are particularly vulnerable have very little chance of getting through the day without being influenced by these triggers, and so the cycle mentioned above repeats itself.

Even when we are not actually hungry, dopamine can be released when we see sugar and the cue–urge–reward cycle is set in motion. New research shows that not only can we form habits easily but also for some people these reward centres can even override hormones that tell us when we are full. This reminds me of my sugar-eating days before I met Zoe. I would be really full after a big meal, but if someone brought out a sticky date pudding, I could always find room for it. (Yes, sticky date pudding, I do still think of you.)

1. CUE (we see the sugary item) 2. URGE (dopamine is released. A primal urge that says 'get that' because you need it for energy.) 3. REWARD (get it, temporarily feel good) 4. HABIT (unaware of this pattern and it becomes the norm or aware of the pattern but its pull is too strong)

'THE BRAIN RUNS ON GLUCOSE, IF THE GLUCOSE LEVEL IS CONSTANTLY GOING UP AND DOWN, ZINGING HIGH AND LOW AND BACK AND FORTH THEN YOUR MENTAL FUNCTION IS JUST UNSTABLE. IF YOUR GLUCOSE LEVEL IS STABLE AND NOT FLUCTUATING THEN YOU HAVE MORE CLARITY.'

THOMAS CAMPBELL, RETIRED NASA PHYSICIST AND AUTHOR OF *MY BIG TOE*

HOW I WAS ON TRACK FOR CARDIOVASCULAR DISEASE

Once we understand how fructose is a fat-making machine, we can appreciate the complications it can lead to.

For a long time we have been told the word to look out for in heart disease is cholesterol. If it is high, you'd better worry because you may not last till morning. However, during my research, one doctor described cholesterol to me as essential for the body, like the nails holding up a wooden house: without them the place would fall down. Another likened cholesterol to firemen at a fire; they may be present at the scene but it doesn't mean they are the problem. Cholesterol is sent to treat a problem, not to cause it.

The misconception for years (and one reason for huge sales of heart medications) has been that there are only two types of cholesterol; good and bad. Well it turns out this is not the case.

We now know that there is good cholesterol, HDL, and *two* types of the one we thought was bad cholesterol, LDL. One of these LDLs is big, fluffy and actually quite friendly, and the other is small, dense and a little intense.

The chance of developing Cardiovascular Disease (CVD) increases when cholesterol interacts with triglycerides and becomes the small and dense type of LDL. Some people might just have the big and fluffy type which is not very dangerous but has been labelled as the bad LDL for years (until the small dense one was discovered).

Due to the high triglycerides in my blood – from the fructose turning to fat in my liver – I was in danger of losing my good cholesterol (HDL) and it becoming the dangerous small, dense LDL cholesterol. In simple terms, extra fat in the blood from the sugar helps create the plaque that can block the arteries.

Today scientists are looking at triglyceride levels more than cholesterol when looking for heart disease because it's the triglycerides that can turn the friendly cholesterol into the damaging type. (The pharmaceutical companies are a little reluctant to come around to this point of view.)

CVD is still the leading cause of death in Australia (2011 statistics). It kills one Australian every 12 minutes and affects one in six people. In the UK, coronary heart disease, a form of CVD, is responsible for one in six deaths of men and one in ten deaths of women. Ask most doctors and they will tell you it is mainly caused by eating fat or smoking cigarettes.

'WE WOULD LOVE TO DO THE DEFINITIVE STUDY BUT IT WOULD MEAN WAITING ANOTHER FIVE YEARS. WHAT POSSIBLE RISK TO PUBLIC HEALTH IS THERE IF PEOPLE STARTED LOWERING THEIR SUGAR CONSUMPTION? OR WE WAIT ANOTHER FIVE YEARS AND OBESITY, TYPE 2 DIABETES AND CVD CONTINUES TO GO UP.'

KIMBER STANHOPE, UNIVERSITY OF CALIFORNIA, DAVIS

EARLY SIGNS OF TYPE 2 DIABETES

Type 2 diabetes is a disease that usually takes a long time to develop. Before I started the experiment, everyone involved certainly didn't expect to see any signs of it in such a short timeframe. The general consensus was that my health was good enough for the body to be able to cope with the sudden influx of sugar.

Over the 60 days, my insulin levels nearly doubled and if I continued my high-sugar diet, it was postulated that in these conditions I could be hyperinsulinaemic or pre-diabetic within six months.

Given the number of people I spoke to who said that my diet during the experiment was almost better than their own, there is cause for concern. Type 2 diabetes is one of the fastest growing diseases in the world and is now estimated to kill someone every six seconds. It is a disease of blood sugar metabolism, yet people with it are still recommended to eat a high carbohydrate diet, which breakes down to sugars. Baffling.

No one knows exactly what role the fatty liver from sugar plays in the development of type 2 diabetes, but scientists like Kimber Stanhope at the University of California, Davis, says that it definitely plays a role. If this is true, then how many medications could be avoided if sugar was removed from the diet?

'SUGAR HASTENS THE DEGRADATION OF ELASTIN AND COLLAGEN, BOTH KEY SKIN PROTEINS. IN OTHER WORDS, IT ACTIVELY AGES YOU.'
DR FREDRIC BRANDT, DERMATOLOGIST

SUGAR AND AGEING: HOW SUGAR ROBBED ME OF MY GLOW!

When I first removed sugar from my diet a few years ago, and again after the experiment, the first thing people would comment on was my skin. Zoe told me that when I was eating sugar my skin looked like it did when I smoked – dry and a bit wrinkly (thanks, honey). Within two weeks of removing sugar, however, my skin apparently 'glowed'.

This is all down to something called glycation, which is triggered when sugar hits the bloodstream. The sugar molecules bind to our protein fibres, especially the collagen and elastin fibres, the building blocks of our skin. When the sugar attacks these fibres, they lose their buoyancy and resilience; the skin then starts to sag and can look old and wrinkly. (There is clearly an area of the male anatomy that the sugar heads straight to first!)

A good overall analogy is to picture a mattress. Imagine the sugar attaching to the coils inside the mattress (the protein fibres). As the coils erode away, the surface of the mattress is going to lose its 'springy' quality, and start to sag and look old. This is what is happening just under the skin when we eat excess sugar. Chocolate biscuit, anyone?

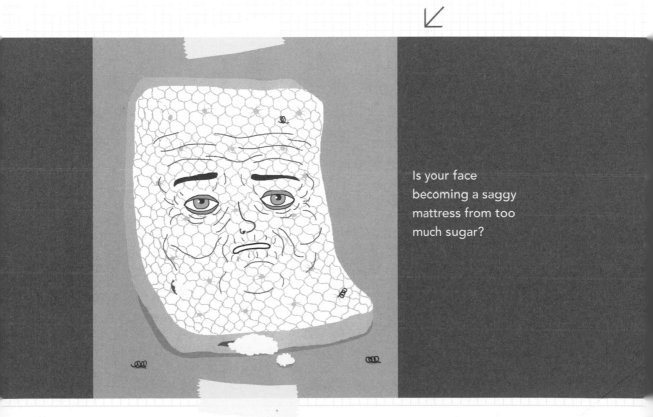

Is your face becoming a saggy mattress from too much sugar?

HOW SUGAR DENTED MY BODY'S DEFENCE SYSTEM

Another surprising result to the team was the dramatic fall in my levels of bilirubin. (This gets my vote for the best-named part of the human body, second only to the vas deferens.) Bilirubin is responsible for the yellow tinge in bruises and the colour of urine. One of bilirubin's roles is as an antioxidant for cells. But what does that mean?

The oxidation in our cells is crucial for life but sometimes this oxidation process can produce free radicals. These can build up steam and eventually lead to cell damage or even cell death. As a result our body needs antioxidants to keep everything in check. These antioxidants are sent to the front line to take down these free radicals and make sure not too much damage is done.

Perhaps it helps to think of the free radicals as terrorist groups and the antioxidants as the UN soldiers trying to keep order in the body. Given that bilirubin acts as one of these antioxidant soldiers, those with high levels of bilirubin are associated with having greater health benefits than those with lower levels – more peacekeeping troops in the body is a good thing.

Now given that my bilirubin levels dropped during the sugar experiment, this implies that my defences were lowered. No one could tell me exactly what this meant but I did find this in a Canadian biochemistry journal: 'Bilirubin, via its antioxidant potential, has a protective effect against cardiovascular disease in young men and women.' (At which point, I was thankful I had ended the sugar-eating lunacy.) Given that I was already in the danger zone for heart disease with my triglyceride levels, it's fair to say I may have needed all the bilirubin troops I could get.

To put it in numbers, my bilirubin levels before the experiment were 18 (with 20 considered to be really healthy). This dropped to six after just 60 days. Precious, antioxidant troops withdrawn from the front line by sugar.

OTHER ILLNESSES

While I thankfully did not experience any of the following illnesses during my experiment, an increasing number of studies have linked sugar with the following:

Gout: A type of arthritis caused by high levels of uric acid, which is now linked to excess fructose consumption. High uric acid levels can also affect the kidneys. Could this be related to the shocking levels of kidney damage in Aboriginal communities?

Alzheimer's disease: People on a high-sugar high-carbohydrate diet are four times more likely to develop Alzheimer's disease. A study from Albany University in New York shows that 70 per cent of people with type 2 diabetes will go on to develop Alzheimer's.

Kidney failure: Linked to type 2 diabetes. Kidneys contain blood vessels that act like filters for the blood. High levels of blood sugar make the kidneys filter too much blood which puts too much pressure on them. The kidneys can start to leak and eventually fail.

Cancers: Sugar molecules can aid the growth of malignant cells. High-sugar diets have also been recently linked to pancreatic cancer.

Candida: These are parasites living in the gut that love sugars and feed off them. They are a type of yeast that, when out of control, can cause fungal infections. Generally, the protocol for getting candida under control is cutting out all sugar from the diet for at least six months. A friend described them as 'rats hanging around rubbish' and said: 'You will only get rid of the rats when you clean up the rubbish.'

Excessive sugar consumption is also linked to hypertension (high blood pressure), tooth decay (Mountain Dew anyone?) and inflammation.

THE GOOD NEWS: AND HOW I GOT HEALTHY AGAIN

PART THREE

'HOW LONG DOES IT TAKE? THERE IS A GRIEF PROCESS FOR SOME PEOPLE AND YOU HIT THE GREY AND THE FLAT, AND THEN ONE DAY YOU MOVE FROM THE GREY TO THE CALM, AND THEN YOU WAKE UP ONE DAY AND SAY, "I REALLY WANT TO LIVE FROM THIS PLACE."'

KATHLEEN DESMAISONS, AUTHOR OF
THE SUGAR ADDICT'S TOTAL RECOVERY PROGRAM

REMOVING THE SUGAR

Okay, I think it's time for some positive news.

After just two months of returning to my normal sugar-free diet, I had lost six of the 8.5 kilograms gained, plus all my blood tests had returned to what they were before I started the experiment. No more fatty liver, no more heart-disease risks and a huge reduction in the visceral fat around my organs – all by just removing the sugar. (Also note I have done minimal exercise as I've been sitting in a chair every day, either editing the film or writing what you are reading at this very moment.)

This quick recovery was a surprise to everyone involved in my experiment and provides great hope. Once I stopped the influx of sugar, my liver stopped making fat and could return to its normal functions; then as my insulin levels declined, the fat was burnt off. I have heard it can be almost impossible to lose abdominal fat unless liver function improves. When it does, the liver can start burning fat efficiently again and the weight will come off.

But removing sugar from your diet can be tricky if you're not prepared. I got through by believing I wasn't giving up anything or 'quitting', I was actually about to gain a healthier way of life and a host of other rewards. We can approach the process with fear or excitement – there is a choice.

I was fortunate enough to have a few very knowledgeable people to help me through the process of coming off sugar. I also had the advantage of having done it before and knowing what I was returning to – I much preferred the sugar-free version of myself. I looked forward to being that person again. For many people however, there is no reference point and there can be a real sense of stepping into the unknown. It can be a daunting prospect: 'What am I going to drink with lunch?', 'What will be my after-dinner reward if I can't have ice-cream?', 'How am I going to live without chocolate biscuits if I'm feeling sad?'

Remember that sugar releases the same beta-endorphins as love, so you may feel like you are going through a break-up when you stop. This is perfectly normal. But a much better love affair lies just around the corner. Sugar was a quick pash in a nightclub but now you're ready for marriage.

An important point to note was that during the experiment I had developed habits. I had been conditioned and my body came to rely on a sugar hit for a temporary, feel-good burst of energy.

In the early 20th century, Ivan Pavlov (who is better known for his Pavlov's Dog experiment) gave a group of volunteers a high-sugar snack for five consecutive mornings. For days afterwards, these volunteers craved something sweet at about the same time each morning, even if they didn't usually eat at that time. This is conditioning. It can be very difficult to break, especially if you have eaten sugar every day of your life since childhood and are attached to the temporary good feelings that come with it. It is only when we become aware of the pattern we can begin to change it.

When I was a smoker, I felt lousy when I went without a cigarette. This only reinforced my need to feel good again, which in my head came from the cigarette, so I lit up another one. Sugar has a similar effect. What really helped me through the early stages of coming off sugar was understanding that the sugar itself was creating the cravings and the ill health. Once I removed the sugar and started eating well, the cravings fell away and the genuine feelings of health and happiness took over. That is the moment when all the fuss about sugar really made sense to me, and you cannot possibly understand that until you have experienced it for yourself.

ORGANIC

APPLE JUICE

NUTRITION INFORMATION

Servings per container: 10
Serving Size: 200mL

Avg Quantity	per Serving	per 100mL
Energy	400kj	200kj
Protein	0.2g	0.1g
Fat, total	0g	0g
- saturated	0g	0g
- trans	0g	0g
- polyunsaturated	0g	0g
- monounsaturated	0g	0g
Carbohydrate	22.5g	11.5g
- sugars	20.4g	10.2g
Dietary Fibre	0.2mg	0.1mg
Sodium	7mg	4mg
Vitamin C	80mg	40mg

* Recommended dietary intake (Australia)

Ingredients: Reconstitued Apple Juice (98%),
Folic Acid, Vitamin C, Flavour MADE IN AUSTRALIA

Shake well before use. Best served chilled.
Refrigerate after opening & consume within 5-7 days.

9 771234 567003

HOW TO READ LABELS

Learning to read food labels really helped me to reduce my sugar intake. A study by Dr Rebecca Huntley revealed we get 55 per cent of our nutritional advice from a packet or label and only 24 per cent from health advocates. So with companies putting profits ahead of health, it pays to know how to read a label carefully. When it comes to understanding how much sugar we are actually consuming, it is vital.

Companies fight very hard to avoid front-of-packet labelling and when you do find the nutrition table, sometimes it can feel like you need a maths degree to figure it all out. The key with sugar is to know that 1 teaspoon = 4.17 grams of sugar (4 grams makes it easier to calculate how many teaspoons of sugar you are consuming over a day).

For example, say you look at the nutrition table on a food label and it states SUGARS 20 g. You can calculate the number of teaspoons in that food by dividing this number by 4 (grams contained in a teaspoon). Therefore there are 5 teaspoons of sugar in the product.

But it gets trickier...

Opposite is the label for a large bottle of apple juice.

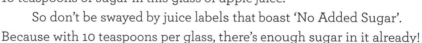

This label tells us that *per serving* this drink contains over 20 grams of sugar. So there are 5 teaspoons of sugar in this serving. But here is where companies get cheeky. This *per serving* size is 200 ml. If we pour 200 ml into a glass we realise what a small serving this is – I would normally have double that at least in one serving. This means we now have 400 ml and we need to double the *per serving* size. So we now have 40 grams or 10 teaspoons of sugar in this glass of apple juice.

200mm   400mm

So don't be swayed by juice labels that boast 'No Added Sugar'. Because with 10 teaspoons per glass, there's enough sugar in it already!

In Australia and the UK, companies do not have to be specific about what type of sugar is in the food. For example, yoghurt may contain lactose plus sucrose but the label will just say 'sugars' (see label on next page).

I calculate that 200 grams (about half a large tub) of yoghurt contains about 9 grams of lactose. According to the label, the yoghurt contains 28 grams of sugar per serving. A serving is 50 grams. So if 200 grams of yoghurt contains 9 grams of lactose, a 50 gram serving will contain 2 grams.

If the total sugar content in a serving is 28 grams, and lactose accounts for just 2 grams, the remaining 26 grams of sugar is likely to be added sucrose. This is just over 6 teaspoons of sucrose per serving.

The American Heart Association recommends no more than 9 teaspoons a day for men and 6 for women, and government advisors in England recommend no more than 7–8 teaspoons for men and 5–6 teaspoons for women. (Australia's Heart Foundation currently doesn't set

LOW FAT

yoghurt BERRY BLISS

INGREDIENTS: Cultured grade A non-fat milk, strawberries, blueberries, water, sugar, fructose, modified corn starch, natural flavour, carrageenan, carmine (for colour), malic acid, potassium Sorbate, (to maintain freshness), Sodium Citrate, Vitamin D3
MADE IN AUSTRALIA

NUTRITION INFORMATION

Servings per container: 10
Serving Size: 50g

Avg Quantity	per Serving	per 100g
Energy	400kj	800kj
Protein	0.2g	0.4g
Fat, total	0g	0g
- saturated	0g	0g
- trans	0g	0g
- polyunsaturated	0g	0g
- monounsaturated	0g	0g
Carbohydrate	22.5g	45g
- sugars	28.0g	56g
Dietary Fibre	0.2mg	0.4mg
Sodium	7mg	14mg
Vitamin C	80mg	160mg

* Recommended dietary intake (Australia)

teaspoon guidelines about sugar intake). When you consider that a glass of apple juice contains 10 teaspoons of sugar, you can appreciate just how high our current sugar intake is, and how damaging it is to our health.

Also, if a food item contains more sugar than anything else it must appear first in the ingredients list. But if a food has a variety of sweeteners, they can be pushed further down the list which can give the impression that there may not be much sugar in the product. This is why sometimes you may see a few different types of sugar contained in the one item: for example, agave, fruit purée, juice concentrate; the company then doesn't have to list 'sugar' first. Sneaky buggers.

Author Michael Moss shared with me a great game he plays with his young kids at the supermarket. He gets them to calculate the amount of sugar in items as they shop. They then try and find the products with the lowest amount. This empowers the kids as they have chosen the foods themselves and feel a sense of achievement – which in turns means they are more inclined to eat their (low-sugar) choices.

CHEEKY HIDDEN SUGARS

Cheeky hidden sugars also lie in the following foods. Read the label and practise the 4 grams equals 1 teaspoon rule.

>**Condiments, such as tomato and barbecue sauce, and salad dressings**
>**Muesli bars and other snacks**
>**Baked beans**
>**Cereals**
>**Tinned soup**
>**Pasta sauce**
>**Tinned fruits**
>**Muffins, cakes and biscuits**
>**Cigarettes** – 15 per cent of a cigarette is sugar.

'ONCE YOU KNOW THE FOOD COMPANIES' TRICKS, YOU CAN BE EMPOWERED AND GET A KICK OUT OF SHOPPING. KNOWING HOW THEY ARE GOING AFTER US IS A GREAT PLAYING-FIELD LEVELLER.'
MICHAEL MOSS, AUTHOR OF *SALT, SUGAR, FAT*

SUGAR IN DISGUISE

The food industry is always finding new and inventive words to use on labels instead of 'sugar'. Professor Robert Lustig has found 56 different names for sugar. These include: Evaporated Cane Juice, Raw Organic Cane Sugar, Fruit Juice Concentrate, Apple Juice Concentrate, Fruit Purée, Cane Syrup, Beet Sugar, Caster Sugar, Crystalline Fructose, Blackstrap Molasses, Grape Sugar, Invert Sugar, Fruit Juice, Turbinado Sugar, Maple Syrup.

When I asked Professor Barry Popkin at the University of North Carolina about these different terms, he replied, 'People need to understand that whether sugar is white, brown, raw, high fructose corn syrup or fruit juice concentrate, all have equal effects on your health.'

BLACKSTRAP MOLASSES

APPLE JUICE CONCENTRATE

GRAPE SUGAR

MAPLE SYRUP

EVAPORATED CANE JUICE

BREAK THE CYCLE

After learning about how the brain works in regards to sugar and dopamine (the primal urge that tells us to get sugar and get it now!), I knew I had to be aware of my triggers and cues in order to break the cycle. My first mission was to empty the house of anything sugary. I didn't want to stumble across a box of Maltesers, the perfectly named temptress, at 3 am.

Getting rid of such cues is vital when you remove sugar from your life. Make the home a safe place for you and your kids, because the food industry has set up 'booby traps' everywhere else. Once you know how it works, it is easier to break the cycle. Go back and read the section on rewards if it helps (see page 122). Any sugary cues left lying around the house can set the cycle of cue–urge–reward–habit in motion. Nip it in the bud before it even gets a chance to start.

In his book *The End of Overeating*, David Kessler has some great advice for those attempting to break eating patterns. This is how it helped me.

AWARENESS

This is the crucial first step when cutting back on sugar – and for most people reading this book, congratulations, you have now acquired awareness. Now you can read a label and work out just how much sugar is contained in your food; you know what it is doing to your mind and body. You are now well equipped to take action if you want to.

Awareness is also key when you face a tricky sugary situation. If you find yourself staring down those Maltesers at 3 am, you can either take a bite, enter the sugar party, feel good for fifteen minutes and then feel crap OR you can use your new-found awareness to observe the dopamine being set off in your brain, laugh at it, know it will pass and then enjoy flushing the Maltesers down the toilet because let's face it, they look like tiny turds anyway.

COMPETING BEHAVIOUR

This is the fun part where you get to shake up life a bit and head in a new direction. It's like Zoe's mum and the airport chocolate shop. Every time she lands in Sydney, she stops by the same shop and picks out a box of delightful treats. She has done the same thing for so many years, her body almost takes her there on auto-pilot. We can all relate to this: it might be an iced coffee from the same deli before work every morning or a bowl of ice-cream after dinner every night. Unless you can consciously override these patterns of behaviour, you will just keep repeating them and the sugar will continue its destructive mission. Zoe's mum now deliberately leaves the airport via a different route to avoid the chocolate shop.

Start by noticing how you shop at the supermarket: which sections do you go to first? You should know that supermarkets are very carefully designed to seduce you into buying certain items. Statistics show that 60–70 per cent of purchases made at a supermarket are unplanned, so we are all giant suckers for the marketing strategies in place. It's time to change the behaviour. Write a shopping list and stick to it. Read the labels carefully; calculate the sugar content. Remember the sugar game that Michael Moss invented for his children (see page 137)? It doesn't have to be all doom and gloom making the change. It might suck a bit at first, but there are fun and creative ways to make it easier for you and your family.

COMPETING THOUGHTS

The idea of sugar as a reward is deeply embedded in our society. We associate it with love, celebrations and commiserations. But these ideas have been brilliantly marketed to us by an industry whose main aim is to make us anticipate reward from their highly palatable products.

Coca-Cola are the masters at it. All their adverts feature attractive people enjoying themselves, high on life and love. Let's not forget that this fizzing black drink can remove rust from bumper bars and contains a huge amount of sugar, which turns to fat in the liver. Our ancient ancestors didn't stand around with Fanta and Gummi Bears at every given celebration; it's not that deeply ingrained.

Every time I felt like a sweet hit in the days after the experiment, I tried to consciously change my thought patterns. I reminded myself that the 'feel good' reward would be temporary and I would soon feel flat or edgy. I also reminded myself of what sugar was doing to the insides of my body.

I found focusing on the positives was really beneficial. I wrote down thoughts about how much better I would feel and look and how much calmer and more present I would be when I could get through the initial phase of wanting sugar and pop out the other side. This last thought was the best for me because it reminded me that I was getting back to a better way of living and Zoe would be happier for it, as would all of my friends and family. I even wrote these thoughts into the notes section of my phone and would refer to it if cravings started creeping in.

I want to add here that it is very important to still have rewards in our lives. It's just time to switch to different ones as the sweet ones are doing us damage. Mine is currently a mug of Bonvit dandelion tea with a splash of almond milk. I have two a day and the nutritionist says that it is great for my liver. I am also into frozen blueberries with a dob of cream every now and then. Zoe's reward used to be gnocchi drenched in cheese but it is now some of her sugar-free homemade chocolate clusters (see page 224 and brace yourself).

We can change our rewards. It just takes a shift in our thought patterns and being gentle and patient with ourselves.

'I'M NOT SAYING LIFE [SHOULD BE] WITHOUT SWEETNESS, BUT LET'S MAKE THAT SWEETNESS COME FROM LIFE AND SENSATIONS – NOT FROM A BOTTLE, A PIECE OF PIE OR ARTIFICIAL SWEETENERS.'

KATHLEEN DESMAISONS,
AUTHOR OF *THE SUGAR ADDICT'S TOTAL RECOVERY PROGRAM*

'WE HAVE TO BE KIND TO OURSELVES WHEN COMING OFF SUGAR. IT'S IMPORTANT TO ASK FOR HELP AND JOIN WITH PEOPLE WHO ARE IN THE SAME SITUATION. NOBODY IS AT FAULT HERE. THIS IS A HUMAN PROBLEM; IT'S PART OF THE HUMAN CONDITION. OUR SUGAR ADDICTIONS HAVE DEVELOPED OVER HUNDREDS OF YEARS AND WE CAN GET TOGETHER TO SOLVE THEM. NONE OF US IS AS SMART AS ALL OF US.'
DAVID WOLFE, NUTRITIONIST

GET SUPPORT

This is perhaps the most important point of all. I hope a time will come when there is no refined sugar in the school canteen, so that children who don't eat sugar won't feel 'different' or as though they are missing out. But this will only occur when we have embraced the sugar issue as a society. We have to come together for our children. We made the mistakes, learnt the lessons and can now pass on our wisdom.

I was lucky because I had people around me who were very support-ive and wanted me to cut back on sugar (I really was a pain in the arse). More and more online sites and groups are popping up where people can get help and feel like they are not alone. These will only grow in the years ahead. People may scoff now if you remove sugar but it is only because they don't understand. And remember that sugar is very addictive so some people will not go down without a fight. They will defend their addiction to the end because it is like a friend or a lover to them.

Things are changing and they are changing very quickly. Sugar will be a big issue in the years to come and the companies will fight hard. My eighteen-year-old cousin recently removed sugar from her diet and looks much healthier as a result. She told me, 'Yes, it was hard, but you just have to do it – for your health, for your life. And now that I look so different, everyone wants to know how I did it! They no longer give me grief about stopping sugar because they can see it worked for me.'

> 'THE GOOD NEWS IS THAT YOU CAN TURN FATTY LIVER AROUND FAST JUST BY TAKING AWAY ALL THOSE DRINKS (SODAS, JUICES AND SPORTS DRINKS). I MEAN WATER IS A GOOD DRINK.'
>
> JEAN-MARC SCHWARZ, TOURO UNIVERSITY, CALIFORNIA

FOODS TO AVOID

From my own experience, the number one enemy is the sugary drink. These include:

Soft drinks with high sugar content – even organic ones.

Fruit juices – whether freshly squeezed, bottled or from a juice bar, it is still a fructose assault on the liver. Green vegetable juices or smoothies are recommended though, with some carrot as the sweetener.

Kids' lunchbox 'fruit boxes' – no matter the claims of 'vitamins' or 'minerals': check the sugar levels!

Energy drinks

Flavoured waters

Sports drinks and most drinks ending in 'ade' – unless you have just run a marathon and need your glycogen 'batteries' restored quickly. These drinks have been brilliantly marketed but are full of sugar.

Flavoured milks – up to 15 teaspoons in the 600 ml sizes.

Tea or coffee with sugar or chai lattes – the powdered form of chai is full of sugar.

'Well, what is left? What do you drink?' I hear you say. I used to have two cans of Coke a day and loved the vanilla flavour the most. Now I am perfectly fine with just water or some sparkling water, with a squeeze of fresh lemon, lime or orange if I am in party mode. Look out.

8 tsp 7 tsp 5 tsp 25 tsp

FOUND TO BE
IN POSSESSION OF
HIGH SUGAR CONTENT

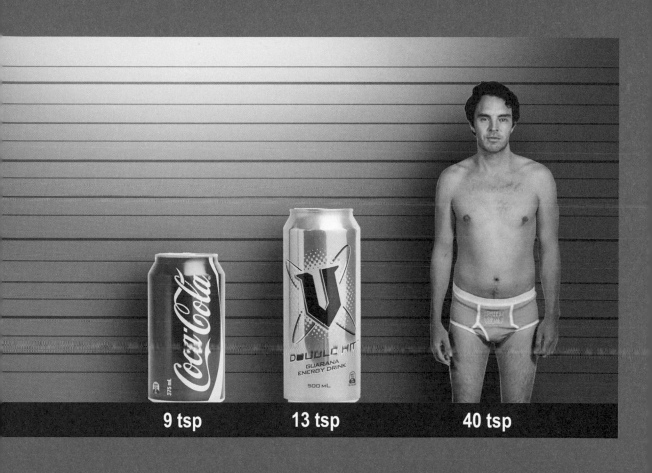

9 tsp   13 tsp   40 tsp

'THIS IS OUR PROBLEM, ALL OF US ON THE PLANET. WE SHOULD ALL WORK TOGETHER. I WOULD LOVE FOR THE FOOD COMPANIES AND SCIENTISTS TO COME TOGETHER AND DO WHAT'S RIGHT – IT'S NOT GOING ANYWHERE.'
ERIC STICE, OREGON RESEARCH INSTITUTE

WATCH THESE FOODS TOO

The following foods spike your blood sugar levels and insulin quickly so can make you feel like eating sugar (sucrose or fructose) when your blood sugars drop again. They also turn off your 'fat burning' processes and make you hold onto fat. I didn't go anywhere near them when I was trying to lose the weight after the experiment.

> White bread, bagels and croissants

> White rice

> Potatoes

> Packaged breakfast cereals and instant oatmeal

> Pasta and noodles

> Crackers, pretzels and corn chips

> Muffins, cakes, muesli bars and pancakes

> Pies

> Pizza

> Beer

> Milk

I avoided all refined carbohydrates: these often contain flour or sugar and are usually white in colour. When sugar is combined with these types of foods then problems can occur.

'I HAD A "WHITE OUT". I SAID I AM NOT GOING TO EAT ANYTHING WHITE: PASTA, RICE, POTATOES AND SUGAR, AND IT WORKED REALLY WELL FOR ME. I AM AWARE THAT I PUT ON WEIGHT VERY QUICKLY WITH CARBOHYDRATES AND THE ONE MOST RESPONSIBLE IS SUGAR.'

STEPHEN FRY, COMEDIAN, ACTOR, WRITER, PRESENTER AND ACTIVIST

AND HOW I GOT HEALTHY AGAIN

MR SUGAR SONG

From the 'Mr Sugar' song in the film (he describes where you might find him):

I'm in cola, granola and pasta sauce

Mushy peas, mac 'n' cheese and that radish from a horse,

Canned fruits and soups, even soup in a satchel

And Mr Sugar loves the term 'All Natural'

White bread, corn chips, muffin mix, gravy,

Mayonnaise, satays, food for the baby

Baked beans for the teens and an energy drink

And I'm clearly going to be in any food that's pink

I'm confessing I'm in dressing and a hamburger bun

Where that cheeky corn syrup's trying to steal all my fun

Muesli bars, even some cigars and a whole range of marmalade
and honey jam jars

A TIP ABOUT LOSING WEIGHT

Understanding the role of glycogen also helped me lose weight quickly. The body gets its energy by first burning off the glucose in our blood; it then burns the stored glucose (glycogen, our 'spare battery') and only after that will it start burning fat. In our sleep we often use glycogen for energy, so any carbohydrates we eat in the morning will go straight to topping up the 'battery' first. Knowing this, doing some light exercise before breakfast, means we would use up the low glycogen levels and then quickly switch to fat-burning mode. Eating toast or cereal before exercise would only top up the glycogen; the exercise would use this for energy first and may not even reach the fat cells. I understand there are lots of opinions on this topic, but this is what works for some people. (A recent study from Northumbria University, published in the *British Journal of Nutrition*, found people lost 20 per cent more fat when not having breakfast before exercise.)

'I feel so bright and bubbly.' That's because you're full of sugar.

WHAT ABOUT ALCOHOL?

For those interested in the role that sugar plays in alcohol: much of the fructose is burnt off in the fermenting process. However sparkling wines, champagnes and dessert wines do retain a lot of their sugar. Beer is made up of the sugar 'maltose' and although it does spike insulin levels which trap fat (higher than white bread), it does not contain fructose. Spirits like gin, vodka and whiskey are low in fructose but it all depends on what you mix them with – go the mineral water option with a squeeze of fresh lemon or lime. Although if you are going to cut out sugar, it is best to avoid alcohol for a while – let your liver get back on track and heal itself.

WHAT ABOUT ARTIFICIAL SWEETENERS?

I have spoken to many people about this topic and the consensus appears to be the same – the body cannot be tricked! A diet soft drink may briefly fool the mind into believing it's a sweet hit, but the body is too smart and will search for its 'legitimate' sugar fix elsewhere. It's likely you'll end up having a muffin or a piece of chocolate later on to satisfy the body's needs.

More to the point, by having artificial sweeteners you are still staying in the realm of sweetness and temptation. What we are talking about here is stepping away from that into a much happier place. Replacements will only serve to keep the feeling and cravings alive. If you are still seeking the sweet hit, even in artificial form, you may find yourself eventually heading back to the full-strength version. Also, trust that your palate will adjust and that you will start to notice the subtle sweetness of more natural foods. (Refined sugar now tastes far too sweet for me.)

Lastly, these diet drinks contain an alarming number of strange sounding chemicals and words with 'x' in them. Respect your body, and look forward to feeling healthy and clean. Your body deserves better than laboratory-made sweetness. Aim higher. If the full-strength sugar hit was a pash in the nightclub and the sugar-free life is a beautiful loving relationship, then the fake sweetener is something really dodgy, probably involving a nightclub toilet cubicle and someone you're not even attracted to.

SNACKS

HUMMUS

HOMEMADE PATE

ALMONDS

MACADAMIA NUTS

HERBAL TEA

WALNUTS

BREAKFAST

BACON

EGGS

AVOCADO

SPINACH

BLUEBERRIES

FULL-FAT YOGHURT

DINNER

COCONUT OIL

STEAK

FISH

CHICKEN

BUTTER

KALE

LUNCH

TUNA

SALAD LEAVES

CHEESE

> 'OUR LOVE OF MEDICATIONS IS LIKE TRYING TO RUN A CAR ON THE WRONG FUEL AND PUTTING AN ADDITIVE IN IT TO MAKE IT RUN WELL. TRYING TO GET ACADEMIA AWAY FROM DRUG TREATMENT BACK TO NUTRITION AND TO USE THE RIGHT FUEL HAS BEEN A VERY HARD TASK. OFTEN THE SIMPLE ANSWERS CAN SEEM BORING, OLD WORLD AND UNSOPHISTICATED, BUT THEY WORK.'
>
> DR SIMON THORNLEY, UNIVERSITY OF AUCKLAND

WHAT I ATE POST EXPERIMENT

I am not going to pretend to be a dietitian or nutritionist. I am just going to share what worked for me. I have learnt that the body is an incredibly intricate system when it comes to dealing with food and what works for some may not be so effective for others. What I did was return to a diet that acted like a 'metabolic reset' button. It was the quickest way for me to reverse my fatty liver, reduce the heart disease risks and shed the dangerous visceral fat. I suspect it may have a similar effect on many others. The following list is for people who want to cut out sugar completely, like I did. Others however may want to begin by just cutting back gently.

My dad, for example, was surprised at how much weight he lost and how much better he felt just by removing a bottle of Gatorade after golf, not mixing Coke with his scotch and having dessert one evening a week instead of five. He now enjoys the feeling and looks great, so has decided to cut back the sugar even more. Gentle steps worked for him.

In the first few days of removing sugar, it was important to prevent my blood sugar levels dropping too low. Also, I didn't go without food for any longer than two hours. If I did, my body would crave sweet to get my blood sugars back up again. I needed to keep my body stable as it went through the process, especially early on.

If you are reducing sugar, remember your body is detoxing. The liver is the key organ in cleaning the body and now it is finally getting some rest, it can heal. You may feel a bit odd initially but this is perfectly normal and is actually a good sign. It can mean that your body is releasing all the toxins and is getting back on track.

The following recommendations worked for me:

→ Drink at least **2 litres of water** a day. This is so important as it helps flush the sugary liver clean.

→ Start the day with a mug of **warm water with a squeeze of lemon**; this also helps the liver with recovery.

→ **Poached eggs with avocado** for breakfast will set you up for the day.

→ Snack on small handfuls of **nuts** and **seeds**. No dried fruit though.

→ **Tuna** is great, especially with a fresh salad.

→ **Chia seeds** are a great source of protein and amino acids (which help with sugar cravings). Soak them in almond or coconut milk and they swell up like a filling porridge.

→ Eat some 'bridging' **fruits** and **veggies**, especially in the first week, to make the transition easier. It is important we are kind to ourselves and go gently in the beginning. Have a small serving of sweet potato (boiled, not roasted) and no more than two pieces of the right kind of fruits a day (any more and you will keep the sugar hit alive). **Blueberries**, **raspberries**, **kiwi fruit** and **honeydew melon** are all low in fructose.

→ **Carrot** or **celery sticks** with hummus are great for cravings (see pages 184–185 for more ideas).

→ **Dandelion tea** is terrific for cleansing the liver (Bonvit's has a nutty flavour and is surprisingly delicious).

→ Add **cinnamon** to your food (studies have shown it can help with insulin sensitivity).

→ Eat **meat** and **fish** if you're an omnivore (these have good amino acids for nuero-transmitters in the brain that help avoid cravings). For more information, see the excellent work of author and clinical psychologist Julia Ross. Julia is a pioneer in food addiction who focuses particularly on sugar and refined carbohydrates.

→ Eat lots of **cruciferous green vegetables** (kale, broccoli, sprouts, spinach, etc.). I often finely chop kale and sprinkle it on a dish or on eggs in the morning.

→ **Supplements** such as 'Matrix Detox' powder or amino acids such as L-Glutamine can help with cravings, but these aren't essential, just a bonus if you can afford them.

My diet is quite high in **healthy fats**. I regularly eat avocado, cheese and coconut products and I love the fat on my meat (quality meat). Nothing helps a sugar craving like some fat – it also keeps my insulin levels down (if eaten without sugar or refined carbohydrates) so I can burn the fat off! (For those wondering about dairy: cream, butter and cheese don't have much of an effect on insulin, but I do avoid milk as that can spike insulin quite dramatically.) I do not fear eating fat and I eat far more of it than I do carbohydrates (see page 235 for suggested reading).

The clinical pathologist who helped take my blood tests during the experiment was so inspired by my results that he decided to cut out sugar and other refined carbohydrates. He has lost 12 kilograms in seven weeks. He said his blood tests are the best they have been in 27 years and he is bursting with a new-found energy. Terrific stuff.

But remember, if you are going to eat more healthy fats, it is imperative that you cut out the sugar and refined carbohydrates so your insulin doesn't spike and trap the fat in your body.

'WHEN YOU WALK INTO A SUPERMARKET MAKE AN IMMEDIATE RIGHT OR LEFT TURN AND HEAD FOR THE PRODUCE AISLES. AVOID THE MIDDLE. ALL THE FOODS IN THE MIDDLE ARE HIGH IN CALORIES WITH LOW NUTRIENTS. WE NEED TO REVERSE THAT.'
DAVID WOLFE, NUTRITIONIST

IN CASE OF EMERGENCY . . .

■ **Remember the sugar itself is creating the craving.** Once you cut out sugar, the cravings will also disappear. It gets easier and easier every time you say no (this really helped me).

■ **Remember the craving is temporary.** If you succumb you will feel flat again in 20 minutes. What you are feeling now will diminish and the brain will begin to rewire itself and heal. THIS MOMENT WILL PASS.

■ **Write a list featuring all the positives of not having the sugar.** Glowing skin, clear eyes, a healthy body inside, lighter energy, a feeling of achievement, more patience with the kids, more genuine laughter in the house, compliments from friends and strangers, more motivation, etc. Write whatever feels inspiring to you.

■ **Write a list of all the things sugar is now linked to.** Fatty liver disease, type 2 diabetes, obesity, heart disease, gout, hypertension, Alzheimer's, mood swings, learning difficulties, dry skin, less brightness in the eyes, puffiness, gut bacteria problems. Do you really want to encourage any of these?

■ **Call a close friend.** Try and see the funny side of the situation and laugh together. The real you is strong and powerful; you may have a small sugary creature on your shoulder trying to tempt you but you are far stronger! There is a great story about a circus elephant with a chain around its leg. If only the elephant realised how powerful it was, it would snap the chain and break free. You are that powerful – sugar is minuscule.

■ **Remind yourself that the sugar isn't a reward at all.** We have been told since childhood that sugary foods are our special treat and are associated with good times. But is something that is doing you damage actually a treat? This really helped me through the cravings (for cigarettes too).

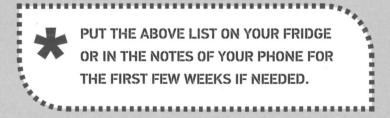

PUT THE ABOVE LIST ON YOUR FRIDGE OR IN THE NOTES OF YOUR PHONE FOR THE FIRST FEW WEEKS IF NEEDED.

EMERGENCY FOODS

Try any of the following foods to help banish cravings:

→ A spoonful of coconut oil

→ A handful of pecans, macadamia nuts, walnuts or almonds

→ A scoop of avocado

→ A sip of apple cider vinegar (foul, but does the job)

→ A warm sweet potato

→ Some chia seeds soaked in almond or coconut milk or water

→ A glass of L-Glutamine powder or tablet (*this helps the neuro-transmitters in the brain to alleviate cravings*).

IN CASE OF EMERGENCY . . .

BREAK GLASS

→ ● ←

READ LIST

I would like to remind you that I am the guy with a certified, scientifically tested sweet tooth of a ten-year-old. I am also the guy who used to drink two vanilla-flavoured Cokes a day and destroy Snickers bars on a regular basis. If I can cut out sugar, anybody can.

THE RECIPES

PART FOUR

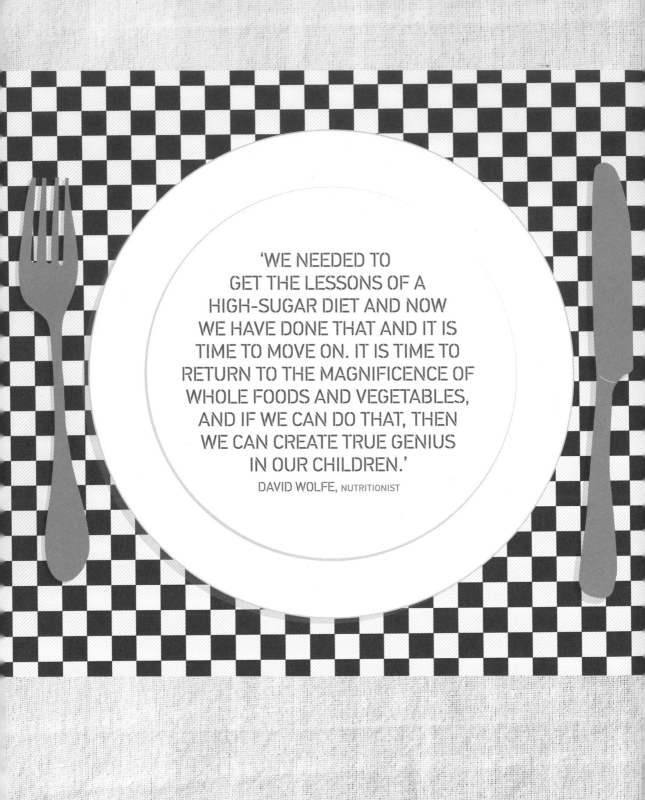

'WE NEEDED TO
GET THE LESSONS OF A
HIGH-SUGAR DIET AND NOW
WE HAVE DONE THAT AND IT IS
TIME TO MOVE ON. IT IS TIME TO
RETURN TO THE MAGNIFICENCE OF
WHOLE FOODS AND VEGETABLES,
AND IF WE CAN DO THAT, THEN
WE CAN CREATE TRUE GENIUS
IN OUR CHILDREN.'

DAVID WOLFE, NUTRITIONIST

This section of the book is a special bonus. I have often loudly proclaimed the culinary skills of my girlfriend, Zoe. She has always loved cooking and does it with a simple ease and flair. It's really exciting for me to be able to grant you an 'all access' pass to our kitchen.

The following meals are some of the ones I ate during my detox after the sugar experiment. Zoe collaborated with recipe developer/writer extraordinaire Michelle Earl, and Sharon Johnston, the nutritionist who helped me throughout the journey. The meals are delicious, nurturing and most importantly, nearly all are devoid of any sugar or refined carbohydrates. This way, my liver could begin to heal and insulin wasn't spiked so the fat could be burnt off. The meals were also designed to help stabilise my moods, making the whole post-sugar period benefical for everybody.

There are a few special recipes for children, including treats. My hope with our own daughter is that her version of a 'special treat' comes from the ones included in the following recipes. Our goal is to protect her from refined sugar for as long as we can, but still give her Zoe's homemade chocolates to establish a new benchmark for what 'sweet' tastes like. The hope is that the refined stuff will then taste too strong for her in the years to come. I can hear most parents sniggering, 'Good luck buddy,' but I can dream can't I?'

We are trying to get both the film and the book into as many schools as possible so we can try and shift the current paradigm. I like to imagine a day when sugary foods won't be waiting for children at recess or lunch in their canteens. Children now get 37 per cent of their daily energy from the tuckshop so we need to be vigilant about the foods they are eating in order to give them the best possible chance to learn.

For anyone thinking about embarking on a new life with a lower sugar intake, I sincerely wish you lots of luck and hope you get as much enjoyment out of the following recipes as I do. And remember, the palate does adjust! Things tasted a little bland to me in the first week, but then sugar's 'numbing effect' on the taste buds began to disappear and I noticed the subtle flavours of healthy foods again. It really is a wonderful moment when this happens and, believe me, it will happen.

For more of Zoe's recipes that helped me off the sugar and allowed me to lose the weight and stabilise my moods, visit: www.thatsugarfilm.com.

Also, visit our dear friend Janneke Williamson at: www.mothernourish.com.

And for my nutritionist, Sharon Johnston, go to www.sharonjohnston.com.au.

THE GREEN PROTEIN MACHINE
VEGETARIAN OMELETTE
WITH BASIL PESTO

3 free-range eggs

1 tablespoon finely chopped mixed herbs, such as flat-leaf parsley, thyme and oregano

10 g butter

Small handful of baby spinach leaves

1 cup (about 125 g) mixed cooked vegetables
or leftover roast vegetables (see Note)

Basil pesto, to serve

This is such a great alternative to cereal. It will keep you satiated well into mid-morning or beyond. But remember, no toast.

Whisk the eggs and herbs in a bowl with a teaspoon of water. Season well with salt and cracked black pepper.

Melt the butter in a small (22 cm) frying pan over medium heat. Add the spinach and vegetables to the pan and cook for a few minutes to warm through and wilt the spinach.

Spread the vegetables evenly in the pan. Pour over the egg mixture and cook for a few minutes until the egg is just set.

Slide out onto a large plate. Serve immediately with a spoonful of store-bought or freshly made basil pesto.

To Make the Basil Pesto
Put 2 cups (50 g) basil leaves in a food processor or blender. Add a small chopped garlic clove, ¼ cup (30 g) pine nuts or almonds, ⅓ cup (30 g) grated parmesan and process to finely chop. With the motor running gradually drizzle in 4–5 tablespoons olive oil until creamy and smooth. Store leftover pesto in a jar covered with extra oil and refrigerate.

Note: *Steam or roast small cubed vegetables such as pumpkin, sweet potato, courgette, spinach, mushrooms, asparagus and especially peas.*

SERVES 1

FRUCT OFF FRITTERS
COURGETTE & PEA FRITTERS WITH SIDE TOMATO SALSA

300 g (2 medium) courgettes

½ cup (75 g) frozen peas, defrosted

2 spring onions, finely chopped

1 tablespoon each finely chopped flat-leaf parsley and mint

2 teaspoons grated lemon zest (ensure you use organic or remove the wax in boiling water before grating)

3 free-range eggs, lightly beaten

¼ cup (25 g) almond meal

1 knob of butter, for frying

TOMATO SALSA
3 Roma or plum tomatoes, seeded and finely chopped

½ small red pepper, seeded and finely chopped

½ spring onion, finely chopped

1 tablespoon capers

1 tablespoon olive oil

2–3 teaspoons red wine vinegar

These are great for school, work lunches and picnics.

Grate the courgettes. To draw out excessive moisture, sprinkle with a liberal amount of salt and set aside for 15 minutes. Then, using your hands, squeeze out all the moisture and place the courgette in a large bowl.

Add the peas, spring onion, herbs, lemon zest, eggs and almond meal. Season well with salt and cracked black pepper.

Heat the butter in a large frying pan over medium heat. Drop large spoonfuls of the mixture into the pan, flatten slightly and cook for 2–3 minutes each side, turning once until cooked through. Cook in 2–3 batches and add a little more butter if needed. Keep the fritters warm until they're all cooked.

Serve immediately with a dollop of the salsa.

To make the Tomato Salsa
Combine the tomato, pepper, spring onion and capers in a bowl. Add the olive oil and vinegar, to taste, season and toss well.

Hint: Replace the peas with corn kernels. Also, replace the Roma tomatoes with a punnet of cherry tomatoes, halved.

Note: Leftover fritters will keep covered and refrigerated for a day. Also good eaten at room temperature.

MAKES 8 FRITTERS
SERVES 4

THE COCOBANADO SMOOTHIE

1 small avocado

1 fresh or frozen banana

About 1 cup (250 ml) coconut milk

1 tablespoon coconut oil

Great for kids as a transition away from sweeter drinks. Very filling, full of energy and contains natural sweetness. It still contains fructose from the banana but remember it's about 'bridging' off sugar early on and easing yourself or your children into a new way of eating. It's important to be kind to ourselves in the first few weeks of any transition.

Put all the ingredients in a blender (add the coconut oil just before you blend so that it doesn't go solid).

Blend until smooth. Thin with more coconut milk or coconut water until you get the consistency you like.

Hint: *Peel the banana before you freeze it.*

MAKES ABOUT 2 CUPS (500 ml)

BERRY POWER SMOOTHIE

1 cup (250 ml) unsweetened almond milk or coconut milk

½ fresh or frozen banana

¼ cup (40 g) frozen berries

1 tablespoon rice or pea protein powder

1 teaspoon raw cacao powder (or cocoa powder if you don't have cacao)

Put all the ingredients in a blender. Blend until smooth. Thin with more almond milk or coconut milk until you get the consistency you like.

MAKES ABOUT 2 CUPS (500 ml)

THE SATURDAY MORNING SPECIAL
SCRAMBLED EGGS WITH SPINACH, GRILLED TOMATOES, BACON & HALOUMI

1 vine-ripened tomato, halved

5 cm slice haloumi, about 1.5 cm thick

5 g butter, melted & extra for eggs

2 small rashers bacon

Large handful of baby spinach leaves

3 teaspoons cream or butter

Pinch of ground nutmeg

2 free-range eggs

1 tablespoon finely chopped herbs, such as flat-leaf parsley, oregano and chives

This is one of my favourites. It's so delicious and has often kept me going until lunchtime and beyond. Just remember to have it without toast.

Preheat a grill. Lightly brush the tomato and haloumi with the butter. Put on a grill tray and add the bacon. Grill for a few minutes, turning the haloumi and bacon once. Cook until the haloumi is hot and bubbly and the bacon is cooked. Keep warm.

Steam or microwave the spinach briefly until wilted. Drain and stir in 1 teaspoon of the cream or butter and the nutmeg. Keep warm.

Whisk together the eggs, herbs and remaining cream. Season well with salt and cracked black pepper. Heat a small frying pan over low heat. Add a knob of butter. Pour in the egg and gently stir until the egg starts to set and becomes creamy.

To serve, spoon the egg onto a serving plate, pile the spinach onto the plate and arrange the tomato, haloumi and bacon on the side.

Remember to have it without toast.

Hint: *Replace the bacon with smoked salmon and the tomato with avocado to make a Sunday Morning Special.*

Note: *You can easily double this recipe to serve 2. Cook the scrambled eggs in a larger frying pan.*

SERVES 1

FOR WHEN YOU'RE FEELING 'EGGED OUT'
THE NAKED BIRCHER

1 small red apple

1 heaped tablespoon chopped nuts, such as almonds, macadamia nuts or Brazil nuts

½ cup (125 ml) full-fat Greek yoghurt

½ teaspoon ground cinnamon

This shouldn't be eaten more than twice a week during the detox period.

Core and grate the apple, including the skin, into a bowl. Add the nuts, yoghurt and cinnamon.

Cover and refrigerate for a couple of hours (or make the night before for an instant breakfast).

Hint: Small amounts of 1 or 2 of the following can be added as extras – frozen blueberries, LSA, chia seeds, sesame seeds, sunflower seeds, pumpkin seeds – not all at once though!

Note: You can easily double this recipe to serve 2.

SERVES 1

CLEAN AND GREEN

A great coffee alternative to start the day. Puts a natural spring in your step.

2 celery stalks

1 small cucumber

Small handful of spinach or kale

½ cup (75 g) chopped pineapple

5 mint leaves, optional

Put all the ingredients in a blender or juicer. Blend until smooth.

SERVES 2

LOVE YOUR LIVER

A nice bridging drink to replace soft drinks or fruit juices. It has subtle sweetness from the beetroot and carrot.

2 celery stalks

1 large carrot or 2 medium carrots

½ medium-sized beetroot

1 teaspoon freshly grated ginger, optional

Put all the ingredients in a blender or juicer. Blend until smooth.

SERVES 2

THE OCCASIONAL AFTER-SCHOOL HOMEWORK HELPERS
NUTTY DATE BALLS

½ cup (175 g) dried pitted dates

1 cup (125 g) walnuts or pecans

1 cup (125 g) macadamias or cashews

1 cup (100 g) desiccated or flaked coconut

1 teaspoon vanilla extract or vanilla bean paste

1 teaspoon ground cinnamon

Desiccated coconut or sesame seeds to roll them in

These do have some fructose from the dried dates, but are a great way to first ease your child off refined sugar. They are a 'bridging' treat.

Put the dates in a bowl and cover with boiling water. Leave to soften for 5 minutes then drain.

Put all the nuts in a food processor and chop them until crumbly. Add the dates, coconut, vanilla and cinnamon. Process until the mixture clumps together in a sticky mass. Bring the mixture together with a splash of water if necessary.

Roll small tablespoons of the mixture into balls (lightly oil your hands to stop them sticking). Roll in coconut or sesame seeds. Store in an airtight container in the refrigerator for 2 weeks or freeze them.

Notes: *You can use any type of nuts for this recipe, whatever your preference. Add a scoop of protein powder if you like. Add 2 tablespoons cacao powder for chocolate version. Blend in 2 tablespoons of chia seeds, linseeds or sunflower seeds with the nuts.*

MAKES ABOUT 20

'I DREAM I GO INTO A GAS STATION ONE DAY AND AT LEAST HALF OF THE FOODS ARE HEALTHY. I ALWAYS THINK THAT WHEN KIDS GO INTO THOSE STORES THEY THINK THAT'S WHAT FOOD LOOKS LIKE, IT ISN'T WHAT FOOD LOOKS LIKE.'

KIMBER STANHOPE, UNIVERSITY OF CALIFORNIA, DAVIS

SCHOOL HOURS BRAIN FOOD
MEAN BEANS & THE VEGGIE DIPPERS

400 g tin cannellini beans (butter beans) or chickpeas, rinsed and drained

1–2 garlic cloves, finely chopped, to taste

2 tablespoons tahini

1–2 teaspoons ground cumin

2 tablespoons each mint and flat-leaf parsley, finely chopped

1 tablespoon rosemary, finely chopped

4 tablespoons olive oil + extra to drizzle

Juice (about 4 tablespoons) and 1 teaspoon grated zest from 1 lemon (ensure you use organic or remove the wax in boiling water before grating)

Paprika, to sprinkle

Carrot, courgette and celery dipper sticks, to serve

It's just so important to not give children sugar or refined carbohydrates at school. It spikes their blood sugar, which can affect their moods and ability to learn or concentrate.

Put the beans in a food processor. Add the garlic, tahini, cumin and herbs. Process to finely chop.

With the motor running, gradually add the olive oil and lemon juice to taste. Season well with salt and cracked black pepper. Adjust seasonings to taste and thin to desired consistency with a little warm water if needed.

Spoon out onto a plate, drizzle with some extra olive oil and sprinkle with paprika. Serve with the vegetable dipper sticks.

Notes: For kids, leave out or add just 1 garlic clove and 1 teaspoon ground cumin for a milder taste.

Spoon into a small airtight container for school/work lunches

MAKES ABOUT 1½ CUPS (425 g)

'DO WHATEVER IT TAKES TO TURN FRUIT AND VEGETABLES INTO THE HIGHLIGHT OF YOUR MEAL AND TRAIN YOUR KIDS TO LOVE THEM, THEY MAY LEAVE FOR THE TEENAGE YEARS, BUT THEY WILL COME BACK AND THEY WILL THANK YOU.'

KIMBER STANHOPE, UNIVERSITY OF CALIFORNIA, DAVIS

SCOTCH EGGS
AKA THE PROTEIN BOMB

4 free-range eggs

4 x 100 g organic beef, pork
or chicken sausages
(or 400 g sausage meat)

1–2 tablespoons cornflour,
for dusting

Coconut oil

Great for school lunches or picnics.

Put the eggs in a saucepan of cold water and bring to the boil. (As they start to boil stir to centre the yolks.) Boil for 5 minutes then transfer to a bowl of cold water. Once they are cool peel them.

Peel the skins from the sausages or divide the sausage meat into four even-sized portions. Put the cornflour on a plate and roll the eggs in it. In the palm of your hand flatten one of the sausage portions into a flat oval (easier to do with damp hands) then mould the sausage meat evenly around an egg. Repeat with the remaining eggs and sausage meat.

Preheat the oven to 200°C (gas 6). Place the eggs on a rack over a roasting tin. Brush lightly with the oil. Cook and turn two or three times for 25–30 minutes until the sausage is cooked and browned. Drain on paper towels. Cut in half to serve.

Notes: If using sausage meat add 1–2 tablespoons chopped fresh herbs and 1 teaspoon tamari. Mix together thoroughly with your hands.

You can also deep-fry or pan-fry the Scotch eggs. If you pan-fry you will need about 3 tablespoons heated oil. Turn frequently until cooked through evenly. Drain on paper towels.

MAKES 4

MORE SCHOOL HOURS BRAIN FOOD
ENERGISING EGG CUPCAKES

10 g butter + extra for greasing

1 spring onion, finely chopped

2 tablespoons chopped red pepper

2 large handfuls of baby spinach leaves

About 6–8 slices thin-cut ham or bacon (see Notes)

4 large free-range eggs

2 cherry tomatoes, halved

2 tablespoons grated cheddar or parmesan

Great fun to make for breakfast and even better for lunch after they've cooled down.

Preheat the oven to 180°C (gas 4). Lightly grease four cups in a small muffin baking tray. In a small frying pan heat the butter over medium heat. Cook the spring onion and pepper for 2–3 minutes until softened.

Wilt the spinach in a steamer or microwave. Cool a little then squeeze out the excess moisture with your hands.

Line the inside of the muffin holes with the ham, enough to completely cover the bases and sides.

Divide the spinach, spring onion and red pepper over the muffin bases.

Carefully crack an egg into each muffin hole. Put a tomato half on the top of each. Season well with salt and cracked black pepper and scatter over the cheese. Bake for 12–15 minutes. (If you are going to carry them for a school or work lunch, let the eggs cook for the longer time to just set the yolks.)

Carefully lift them from the muffin tin and eat straight away or leave them to cool on a wire rack. Refrigerate until needed. Will keep refrigerated for a day.

Notes: *If using bacon, lightly pan-fry or grill it for a couple of minutes.*

Wrap in foil for carrying in a lunchbox.

MAKES 4

LIGHT LUNCH
TENDER CHICKEN VEGGIE WRAPS

8 chicken tenderloins

½ cup (50 g) almond meal

Zest and juice of 1 small lemon (ensure you use organic or remove the wax in boiling water before grating)

⅔ cup (60 g) firmly packed finely grated parmesan or pecorino

1 free-range egg, beaten

1 avocado

4 large iceberg lettuce leaves or rice paper

Shredded vegetables, such as carrot, beetroot, celery, courgette

Sliced cos or iceberg lettuce

1 cup (115 g) grated cheddar cheese

Easy to assemble and an excellent lunchbox idea.

Preheat the oven to 180°C (gas 4). Line a baking tray with baking paper. Pat the tenderloins dry with paper towels.

In a medium-sized bowl mix the almond meal, lemon zest and parmesan. Dip the chicken into the beaten egg and then coat in the almond crumb. Put on the tray. Cook in the oven for 12–15 minutes turning once, until the crust has browned and the chicken is cooked. Roughly mash the avocado with 1–2 teaspoons lemon juice.

To assemble, smear each lettuce leaf with the avocado. Top with two tenderloins, some shredded vegetables and sliced lettuce and top with the grated cheese. Wrap up firmly in foil and pop into a lunchbox.

SERVES 4

WHITE BUTTER BAKED BEANS
WITH ROSEMARY

20 g butter

1 tablespoon chopped rosemary leaves

400 g tin white beans, any variety will do

Large pinch of salt

So simple and delicious. Great as a breakfast side or a snack during the day.

Melt the butter in a frying pan over medium heat. Add the rosemary and then the beans. Allow them to cook in the butter for a couple of minutes.

Add the salt, then pour in water to just cover the beans and let them simmer until the water boils down, stirring occasionally with a wooden spoon to break up the beans so some are mushy and some are still whole.

Serve simply with a little pepper. Or as a breakfast option serve with a poached egg on top and some avocado on the side.

SERVES 2

HAPPY HEART
COCONUT CHICKEN PÂTÉ

500 g chicken livers

Salt

100 g butter, + extra to cook the livers

3 tablespoons coconut oil

1 tablespoon apple cider vinegar

Eat with carrot and cucumber sticks rather than crackers.

Melt a knob of butter in a small (22 cm) frying pan over medium heat. Add the chicken livers, salt lightly and cook until pink in the middle.

Place the livers in a blender, add the butter and coconut oil.

The heat of the livers should melt the butter down.

Add the apple cider vinegar, blend and pour into a bowl. Cover and put in the refrigerator.

Hint: *To prevent the pâté oxidising quickly, put it in a stainless steel lunchbox with a sealed lid.*

MAKES A GREAT SIZE TO KEEP IN THE REFRIGERATOR AND USE OVER THE WEEK

AVOCADO DIP

Mash 2 avocados in a bowl. Add 1 seeded and chopped tomato, 1–2 tablespoons finely chopped red onion, 1 crushed garlic clove, 1–2 tablespoons lemon or lime juice and season well with salt and cracked black pepper. Serve with vegetable sticks.

TZATZIKI

Halve and seed 2 cucumbers, then grate into a sieve over a bowl. Sprinkle with a little salt and set aside for 10 minutes. Rinse under cold water and drain. Squeeze out the excess liquid, using your hands, into a bowl. Stir in 375 g thick Greek yoghurt, 1 small handful of finely chopped mint leaves, 1–2 crushed garlic cloves and 1–2 tablespoons lemon juice, to taste. Spoon into a bowl and serve with vegetable sticks to dip.

QUICK SNACKS

The simplicity of a handful of almonds, walnuts or macadamia nuts. So good when you're having sugar cravings.

Celery sticks with almond butter.

THE MEAL THAT FEELS LIKE A HUG
COMFORTING KALE & CASHEW SOUP

10 g butter

2 garlic cloves, crushed

1 large onion, finely chopped

1 carrot, finely chopped

2 celery stalks, finely chopped

1 small sweet potato, peeled
and cut into 1 cm chunks

2 large handfuls of kale,
de-stemmed by ripping off the
leaves and roughly chopped
(you can also use spinach)

1 cup (150 g) cashew nuts, soaked
in water for at least 1 hour

3 cups (675 ml) vegetable stock
or water

1 tablespoon apple cider vinegar

So nurturing if you're trying to cut back on sugar.

In a large saucepan heat the butter over medium heat. Sauté the garlic, onion, carrot, celery and sweet potato for 8–10 minutes until softened.

Add the remaining ingredients and bring to the boil, then reduce the heat to low and simmer, covered, for 8–10 minutes until the vegetables are well cooked.

Remove from the heat, allow to cool a little then blend until smooth. Thin with extra stock or water, if you like, and season well with salt and cracked black pepper. Reheat gently to serve.

Notes: *Add a dollop of sour cream or yoghurt. Grate in ½ cup (60 g) strong cheddar just before serving. Sprinkle with fresh chopped chives.*

Hints: *To speed up the process chop the onion, carrot and celery into 3 or 4 pieces and pulse in a food processor until well chopped. A high-powered blender will give the best result, but if you don't have one, then take the time to blend with a hand/stick blender for a few minutes.*

Freeze in portions.

SERVES 4

THE HOMEWORK HELPER
AFTER-SCHOOL ENERGY SOUP

1 tablespoon olive oil, +
3 tablespoons extra

2 garlic cloves, finely chopped

400 g tin diced tomatoes

200 ml vegetable stock or water

1 teaspoon dried herbs, such as
rosemary, thyme, oregano and
basil

300 g tin white beans or
chickpeas, rinsed and drained

This soup can be served with seed crackers.

In a medium saucepan heat the olive oil over low heat, add the garlic and sauté briefly until lightly coloured (not browned).

Add the tomatoes, vegetable stock and herbs, increase the heat and bring just up to the boil. Then immediately reduce the heat and simmer for 10 minutes.

Allow to cool a little, then add the extra olive oil and blitz with a hand/ stick blender.

Add the beans, season with salt and cracked black pepper and heat gently for 1–2 minutes.

Notes: *Stir through or sprinkle with fresh chopped thyme, oregano or basil – whichever you used dried. Sprinkle over some grated parmesan, to serve.*

Hint: *Freeze in portions.*

SERVES 4

THE HEALING GREEN BOWL
'I CAN'T BELIEVE IT'S BROCCOLI' SOUP

20 g butter

1 small leek, white part only,
thinly sliced

1 celery stalk, finely chopped

1 garlic clove, chopped

2 broccoli heads, roughly
chopped

3 cups (675 ml) vegetable stock
or water

Handful each of flat-leaf parsley
and mint, roughly chopped

2–3 tablespoons cream, to taste

My stepmother used this recipe for years, tricking me into eating broccoli.

In a large saucepan heat the butter over medium heat. Sauté the leek, celery and garlic for 4–5 minutes until softened.

Add the broccoli, stock and herbs. Bring to the boil, then reduce the heat to low. Cover and simmer for 10 minutes or until the broccoli is tender.

Remove from the heat, allow to cool a little, then blitz with a handstick blender or in a food processor, until smooth. Blend in the cream to taste and season well with salt and cracked black pepper. Reheat gently to serve.

Hints: This makes a thick soup, thin it if you like with extra vegetable stock or water.

A high-powered blender will give the best result, if you don't have one then take the time to blend with a hand/stick blender for a few minutes.

Freeze in portions.

SERVES 4

COOL ALMOND, CUCUMBER & TOMATO SOUP

750 g vine-ripened organic tomatoes, peeled, seeded and cored

1 small red onion, chopped

½ red or green pepper, seeded and chopped

½ cucumber, peeled, seeded and chopped

1 tablespoon capers, rinsed and drained

1–2 garlic cloves, chopped, to taste

1 red chilli, seeded and chopped, optional

½ cup (50 g) almond meal, lightly toasted

2 cups tomato juice

1 tablespoon olive oil

1 tablespoon red wine vinegar, or to taste

Small handful each of fresh coriander and mint, chopped

So fresh and crisp this will have you feeling bright, alert and focused. Great on a warm day too.

Place the tomatoes, onion, pepper, cucumber, capers, garlic and chilli, if using, in a food processor or blender and purée until smooth. Blend in the almond meal.

Transfer to a large bowl and stir in the tomato juice, olive oil, vinegar, to taste, and season well with salt and cracked black pepper.

Cover and chill for at least 2 hours, then stir in the herbs. Spoon into bowls and top with the extra vegetables (see garnish ideas below) as a garnish and for extra crunch.

To serve, finely chop the remaining half of the cucumber and pepper and spoon on top.

Extra garnish ideas

Add a spoonful of baby capers or a chopped gherkin to the vegetable garnish. Scatter over some toasted slivered almonds.

SERVES 4

THE HEALTHY COMFORT FOOD RANGE
THE HUNTER-GATHERER FRITTATA

Coconut oil

10 g butter

2 rashers bacon, rind removed and chopped

1 small leek, white part only, thinly sliced

1 cup (225 g) cubed pumpkin, steamed

2 courgettes, cubed and steamed

About 10 small broccoli florets, steamed

8 cherry tomatoes, halved

Large handful of baby spinach leaves

8 free-range eggs, lightly beaten

1 cup (240 ml) cream or milk

½ cup (115 g) crumbled feta

½ cup (60 g) grated cheddar

Green salad, to serve

A crustless quiche, minus the refined carbohydrates from pastry.

Preheat the oven to 180°C (gas 4). Lightly grease a 25 x 20 x 5 cm baking dish with coconut oil. Heat the butter in a medium frying pan over medium heat. Sauté the bacon and leek until soft, about 5 minutes.

Add all the vegetables to the bacon and leek and gently stir to combine. Place a lid on the frying pan and leave for a couple of minutes until the vegetables have softened.

Combine the eggs and cream in a bowl and season well with sea salt and cracked black pepper. Take the bacon and vegetable mix and tip into the baking dish. Scatter the feta and cheddar over the mixture. Pour the egg and cream mixture over the top. Bake for 35–40 minutes until just set. Serve with a simple green salad.

Notes: *Delicious cold. Cover and keep refrigerated for 1-2 days. Works for a children's lunchbox the next day too.*

Any combination of vegetables can be used. Try left over roasted vegetables, asparagus and mushrooms.

Hint: *Run the coconut oil bottle under a hot tap if it is cold.*

SERVES 4–6

THE DUVET DESTROYER
PUMPKIN & CHICKPEA COCONUT CURRY

1 tablespoon coconut oil

1 large brown onion,
finely chopped

2 garlic cloves, finely chopped

1 tablespoon freshly
grated ginger

1 teaspoon curry powder

2 kaffir lime leaves (available
from Asian grocers, can be
stored in the freezer)

1 cup (225 ml) vegetable stock

½ small pumpkin or squash, cut
into small chunks (about 400 g)

400 g tin chickpeas, rinsed
and drained

Large handful of baby spinach
leaves

150 ml coconut cream

Juice of 1 lime or ½ lemon

This is a beauty from nutritionist Sharon Johnston, aka the Celtic Food Queen.

Heat the coconut oil in a large saucepan over low heat. Add the onion and garlic and gently sauté for a few minutes until soft.

Add the ginger, curry powder and twisted lime leaves (this helps to release the flavour) and stir for 1–2 minutes, then add the vegetable stock and pumpkin pieces. Bring to the boil, then lower the heat. Cover and gently simmer until the pumpkin is soft, around 10 minutes.

Add the chickpeas and cook for another 2 minutes. Add the spinach, coconut cream and lime juice. Check the balance of flavours and season further with lime juice and salt as required.

Hints: This is a mild-flavoured curry, if you prefer you can add a little more curry powder to taste. If you're feeling really brave you can use chilli powder and cayenne as well as curry powder – just to add a little bit of kick! Rather than chilli powder you could use a tiny bit of fresh chopped chilli in a bowl to serve for people to add themselves.

SERVES 2

HEALTHY AND SIMPLE
THE ROCK SOLID FISH DISH

2 handfuls of green beans, topped and tailed

2 carrots, cut into batons

2 handfuls of sliced red cabbage

¾ cup (165 ml) olive oil, + extra for greasing

½ cup (125 ml) Bragg Liquid Aminos or tamari

2 large Roma or plum tomatoes

2 pieces boneless salmon fillet (about 180 g each)

1 lemon, sliced into 6 rounds (keep ends to serve)

From Zoe: 'A light, yet filling meal for those nights when you need protein, but don't want to go the whole hog (so to speak)!'

Preheat the oven to 180°C (gas 4). Cut four squares of baking paper or two squares of foil – large enough to generously enclose the fish in two parcels. Lightly grease with some olive oil.

Arrange the green beans, carrot and cabbage into two piles side by side on the baking paper or foil. Drizzle with olive oil and Bragg Liquid Aminos or tamari.

Slice the tomatoes into rounds and lay 3 slices onto each of the vegetable piles. Place the fish on top and drizzle again with the Braggs and olive oil. Arrange the lemon rounds on top of the fish. Drizzle with the remaining oil.

Fold up the parcels neatly and arrange side by side on the baking tray. Bake for 15–20 minutes until the fish is cooked to your liking (longer for larger thicker pieces of salmon).

Serve fish with additional olive oil, Braggs and lemon pieces on the side if you wish.

Variation: Try other vegetables such as courgettes, fennel or broccoli.

Note: Bragg Liquid Aminos is a delicious alternative to soy sauce and can be found in health food stores as well as supermarkets.

SERVES 2

THE INSULIN-FRIENDLY LASAGNE
AUBERGINE LASAGNE

1 large sweet potato, peeled

1 red pepper, halved and seeded

3 courgettes

6 mushrooms, quartered

Coconut oil

2 large aubergines, thickly sliced

2 free-range eggs

1 cup (240 ml) cream

1 cup (250 g) ricotta cheese

½ cup (60 g) grated cheddar, + ½ cup (60 g) extra

½ cup (50 g) grated parmesan

1 teaspoon ground nutmeg

1 cup (260 g) tomato passata

By using aubergine, this is a terrific alternative to regular lasagne – a great comfort food.

Preheat the oven to 180°C (gas 4). Chop the sweet potato, pepper and courgettes into 3 cm cubes. Spread the sweet potato on a baking tray lined with baking paper. Scatter the pepper, courgettes and mushrooms onto another large baking tray lined with baking paper. Toss the vegetables with a little coconut oil to lightly coat. Bake for 30 minutes turning them once or twice until cooked (the sweet potato will need 5–10 minutes longer). Set aside.

Preheat the grill, brush the aubergine slices with oil and lay slices in a single layer on the grill tray. Grill and turn the aubergine for about 8 minutes until softened. Set aside.

Beat together the eggs, cream, cheeses and nutmeg in a bowl.

Layer into a 25 x 20 x 5 cm baking dish, half of the tomato passata, a layer of aubergine slices, half the roast vegetables and half the cheese mixture.

Repeat with another layer, ending with the cheese mixture. Scatter over the extra cheddar. Bake for 35–40 minutes until the sauce is set and is hot and bubbly. Set aside for 10 minutes then cut into large slices.

Hint: When grilling the aubergine, you may have to do a few batches to get the right amount. The aubergine is the substitute for the pasta and a couple of layers are preferable to get the same 'feel' to the dish.

Tip: Also delicious with beef mince (added in layers).

Another Tip: If you are a pasta lover, try buying a spiraliser (from any kitchen goods stockist) and use it to turn vegetables such as courgettes and carrot into pasta substitutes. It's fun to do with children too. Use the raw vegetable pile as a base for a delicious Bolognese sauce or pesto topping.

SERVES 4–6

NUTTY FISH SCHNITZEL
MACADAMIA NUT-CRUSTED FISH

Coconut oil

4 fillets of boneless meaty white fish, such as snapper or perch (allow 150 g per person)

½ cup (75 g) unsalted macadamia nuts or cashews

Zest and juice of 1 small lemon (ensure you use organic or remove the wax in boiling water before grating)

⅔ cup (60 g) firmly packed finely grated parmesan or pecorino

1 small free-range egg, beaten

Baby spinach leaf salad, to serve

So light, so tasty and so healthy.

Preheat the oven to 180°C (gas 4). Brush a baking tray lined with foil with coconut oil. Pat the fish dry with paper towels.

Finely chop the nuts in a food processor. In a medium-sized bowl mix the nuts, lemon zest and parmesan. Dip the fish fillets into the beaten egg and then coat in the nut crumb.

Gently position the fish on the tray. Cook in the oven for 12–15 minutes, until the crust has browned and the fish is cooked. Splash over the lemon juice and serve with a spinach salad.

SERVES 4

GET COMFY WITH THE
FIRESIDE BEEF STEW

2 tablespoons olive or coconut oil
800 g beef, cut into bite-sized chunks (use chuck, stewing or blade steak)
1 rasher bacon, chopped
4 spring onions, thickly sliced
2 garlic cloves, chopped
1 large carrot, roughly chopped
1 celery stalk, thickly sliced
2 turnips, peeled and cut into large chunks
2 parsnips, peeled and cut into large chunks
400 g tin diced tomatoes
1 cup (225 ml) beef stock or water
1 tablespoon tomato paste
1 teaspoon mixed dried herbs or dried oregano
¼ (60 ml) cup cream
125 g green beans, halved and steamed
Salad, to serve

A hearty meal without any refined carbohydrates or sugar.

Preheat the oven to 130°C (gas ½). Heat half the olive oil in a large flameproof casserole dish over high heat. Sear the beef in 2–3 batches, turning until nicely browned. Remove to a side plate.

Lower the heat to medium. Heat the remaining oil. Sauté the bacon, spring onion, garlic, carrot and celery for a few minutes.

Return the beef to the dish and add the turnip and parsnip. Stir through the tomatoes, stock, tomato paste and herbs. Season well with salt and cracked black pepper.

Cover with a lid and put in the oven for 3 hours, stirring occasionally. To serve stir in the cream and steamed green beans. Serve with a salad.

Hint: *You can also add the beans to the stew for all of the cooking time. They will be very soft. If you like your beans crisp and green, add them lightly steamed at the end as in above method.*

SERVES 4

NEVER LETS YOU DOWN ON A WINTER'S NIGHT
THE NAKED SHEPHERD

2 tablespoons coconut oil

2 garlic cloves, chopped

1 onion, finely chopped

1 carrot, finely chopped

1 celery stalk, finely chopped

1 bay leaf, crushed

600 g beef mince

400 g tin diced tomatoes

1 tablespoon tomato paste

1 tablespoon tamari

¼ cup (60 ml) beef stock or water

1 cup (225 g) fresh or frozen peas

CAULIFLOWER AND CHEESE MASH

1 small cauliflower, cut into florets (about 600 g)

20 g butter

¼ (60 ml) cup cream

1 cup (115 g) grated cheddar (or ½ cheddar / ½ parmesan)

Green salad, to serve

Like Shepherd's Pie but without potato or pastry.

Heat the coconut oil in a large deep frying pan over medium heat. Sauté the garlic, onion, carrot, celery and bay leaf for 5 minutes until the vegetables are softened.

Increase the heat to high and add the meat. Stir and break up any lumps for 5 minutes until the meat is browned. Stir in the tomatoes, tomato paste, tamari and beef stock. Cover and cook over medium heat for 20 minutes, stirring once or twice. Remove the bay leaf. Stir in the peas.

For the Cauliflower Mash, steam the cauliflower florets for 5–7 minutes until tender. Transfer to a blender or food processor with the butter, cream and half the cheese. Season well with salt and cracked black pepper and blend until smooth.

Preheat the oven to 200°C (gas 6). Pile the beef evenly into a deep baking dish. Spoon over the mash, and mark the surface with a fork. Scatter over the remaining cheese. Bake for 30 minutes until the top is golden brown. Serve hot with a salad on the side.

Hint: Halve or quarter the onion, carrot and celery. Pile into a food processor and pulse a few times to finely chop.

SERVES 4

RONALD MC WHO?
NAKED NUGGETS
WITH FANCY FRIES

500 g chicken breast fillets

1 cup coconut flour

2 free-range eggs

1½ cups (150 g) shredded coconut

1 bunch kale

1 teaspoon coconut oil

A great meal for kids and delicious the next day in your lunchbox.

Preheat the oven to 180°C (gas 4). Cut the chicken into bite-sized chunks. Roll in the coconut flour. Beat the eggs on a plate and dip the rolled chicken into the beaten egg. Coat the chicken in the shredded coconut. Put the chicken on a baking tray. Place in the oven.

Tear the curly kale off the stem in medium-sized pieces. Pour the coconut oil into a bowl and toss the kale with the oil along with a sprinkling of salt. Spread the kale on a baking tray.

After approximately 15 minutes move the chicken onto the bottom shelf of the oven to keep gently cooking and place the kale on the top shelf. Turn the oven down to about 130–160°C (gas ½–gas 3), depending on the heat of your oven.

Allow the kale to slowly and gently cook for 10 minutes or so – this is a fine art and you may need to keep an eye on it. You want the edges starting to brown lightly but the chip still green.

Once the kale chips are crispy remove the chicken and kale from the oven. Lightly sprinkle the chicken with salt.

Notes: *You can use thin slices of sweet potato or beetroot for the fries (using a potato peeler). You may need to adjust the oven and cooking time for these variations. The main thing is to keep the coconut oil to a minimum in order to keep the chips crispy and light. Experiment and have fun with it.*

Only salt chicken after cooking as doing it beforehand will dehydrate the chicken and make it less juicy!

SERVES 2–3

SIMPLICITY WITH A TWIST
COCONUT ROAST CHICKEN

1.5 kg whole organic chicken

2 spring onions, finely chopped

1-2 garlic cloves, finely chopped

¼ cup (60 g) almonds, toasted and finely chopped

2 tablespoons almond meal

2 tablespoons desiccated coconut

1 tablespoon freshly grated ginger

½ small lemon or lime, thinly sliced or finely chopped

20 g butter

1 tablespoon coconut oil

Salad or steamed vegetables, to serve

There's something so right about the flavour of coconut with chicken.

Preheat the oven to 200°C (gas 6). Wash the chicken inside and out and pat dry thoroughly with paper towels.

Combine the spring onion, garlic, chopped almonds, almond meal, coconut and ginger in a bowl. Add the lemon, season with salt and cracked black pepper and combine well. Lightly fry in a frying pan in the butter until brown and warm. Stuff the mixture into the cavity.

Tie the drumsticks together, tuck under the wingtips and place in a roasting tin. Rub the coconut oil all over the chicken.

Roast for 15 minutes then reduce the oven to 180°C (gas 4). Baste once or twice with the pan juices. Cook for a further 45–50 minutes. The chicken is cooked when the juices run clear when skewered into the thickest part of the breast.

Cover loosely with foil and leave to rest for 15 minutes. Cut the chicken into pieces (easiest to do with scissors). Plate up with a spoonful of the stuffing and some of the pan juices. Serve with salad or steamed vegetables.

Note: *This can be served with a simple fresh salad of leaves or a side of steamed broccoli for something warmer in winter!*

SERVES 4

THE STAY-AT-HOME PUB MEAL
ALMOND CRUSTED CHICKEN SCHNITZEL

½ cup (50 g) almond meal

1 tablespoon finely chopped flat-leaf parsley

Zest and juice of ½ lemon (ensure you use organic or remove the wax in boiling water before grating)

1 large chicken breast fillet, slice in half widthways to make 2 thinner pieces

1 tablespoon arrowroot or cornflour

1 free-range egg, beaten

2 tablespoons coconut oil

Steamed broccoli and mashed sweet potatoes, to serve

More guilt-free yummy comfort food.

In a shallow bowl, combine the almond meal, parsley and lemon zest.

Lightly coat the chicken breast in the arrowroot then dip into the egg followed by the almond meal mix.

Heat the coconut oil in a medium frying pan. Cook over low to medium heat for 5–6 minutes, turning the chicken once, until lightly golden and cooked through. (Don't have the heat too high otherwise the almond crumb will brown too much.)

Serve alongside steamed broccoli and mashed sweet potatoes, add a squeeze of lemon juice and enjoy.

Hint: If not eaten immediately refrigerate and use for a wrap or salad the following day.

Note: This is one of the 'bridging' foods that will help with slowly coming off a high-sugar diet. It shouldn't be eaten regularly but can be a real asset in the first couple of weeks of reducing sugar.

Variation: Serve with other steamed green vegetables, such as beans and courgettes.

SERVES 2

THE FLOWER OF LIFE
BRUSSELS SPROUT SALAD

500 g brussels sprouts

2 tablespoons olive oil

1 tablespoon apple cider vinegar

1 teaspoon Dijon mustard

Salt and pepper

Dash of soy sauce or tamari

½ cup (75 g) pecans, lightly toasted and roughly chopped

1 spring onion, thinly sliced

Large handful of dried cranberries

1 red apple, cored and grated

60 g goat's cheese

A modern version of the Waldorf salad. Those who have been traumatised by Grandma's overcooked sprouts will be surprised at how delicious they taste raw. The beautiful ancient symbol of the flower of life is found when cutting the sprouts crossways.

Holding each sprout by the stem end, cut into very thin slices using a mandolin slicer or thinly slice with a knife. Toss in a bowl to separate the layers.

In a small jar, shake together the olive oil, apple cider vinegar, Dijon mustard and salt and cracked black pepper.

Just prior to serving add a dash of the soy to the pecans and toss. Add the pecans, spring onion, cranberries, freshly grated apple and small blobs of goat's cheese to the bowl. Pour the dressing over the salad and toss thoroughly.

Hint: *Brussels sprouts can be sliced a couple of hours ahead and chilled, covered in the bowl. Toss with the remaining ingredients just before serving.*

Note: *Replace the goat's cheese with ½ cup (50 g) grated parmesan.*

SERVES 4

Note the beautiful geometry when a brussels sprout is cut in half. A clear winner if there was a vegetable 'make over' show.

A BONA FIDE WINNER
THE BRIDGING SALAD

1 bunch kale

2 tablespoons olive oil

Sea salt

3 teaspoons apple cider vinegar

1 small red onion, thinly sliced

Black pepper

Handful of almonds (about 16) with skin on (preferably soaked in water overnight), chopped

½ cup (50 g) crumbled feta

1 cup (150 g) blueberries

400 g tin lentils, rinsed and drained

¼ cup (25 g) desiccated coconut, lightly toasted

Light and good for transitioning off sugar as it has a sweetness from the blueberries.

Remove the stalks and tough seams from the kale. Tightly roll up the kale leaves and slice into thin ribbons. Put in a large bowl and drizzle with 1 tablespoon of the olive oil and sprinkle with some sea salt. Massage the kale for a few minutes with your hands to soften it, then leave it to sit for 5–10 minutes.

Add the remaining olive oil, vinegar, onion and some freshly ground black pepper. Leave to marinate for 10 minutes.

Just prior to serving toss through the almonds, feta, blueberries, lentils and coconut.

Note: *Replace blueberries with sliced orange and feta with goat's cheese.*

Hint: *Use brown or green lentils. The tinned variety is convenient but you can cook 1 cup (240 g) of Puy lentils in plenty of boiling water for 15 minutes, or until soft, then drain and cool.*

SERVES 4

THE LIGHT LUNCH
CLARITY COLESLAW AND KICKING TUNA

MARINADE
4 tablespoons coconut oil

3 tablespoons tamari

3 tablespoons fresh lemon juice

1 garlic clove, crushed

½ teaspoon freshly ground black pepper

¼ teaspoon dry mustard

2 tuna steaks (about 130 g each)

COLESLAW
½ small wombok (Chinese cabbage) or red cabbage, finely shredded

1 large carrot, peeled and grated

1 small red onion, halved and thinly sliced

100 g mangetout, thinly sliced lengthways

1 cup (40 g) chopped fresh coriander

3 tablespoons tamari

1 tablespoon sesame oil

1 tablespoon finely shredded fresh ginger

1 garlic clove, crushed

Lemon wedges, to serve

Another offering from the Celtic Food Queen.

Prepare the tuna marinade by mixing all the ingredients together in a small jug or bowl.

Place the tuna steaks in a shallow dish and cover with the marinade, refrigerate for at least 1 hour.

Prepare the coleslaw by combining the wombok (Chinese cabbage), carrot, onion, mangetout and coriander in a large bowl.

To make the dressing, whisk together the tamari, sesame oil, ginger and garlic in a small jug. Drizzle over the coleslaw and toss to combine.

Once the tuna has been marinated, place on a preheated non-stick grilling rack and grill for 2–3 minutes on each side or until opaque and still slightly pink in the centre. Brush with the marinade while cooking to keep moist.

Serve with the lemon wedges and the coleslaw.

Hint: For a super-quick version this can also be made with tinned tuna or salmon mixed into the coleslaw.

SERVES 2

BERRYBANANA ICE BLOCKS

Use raspberries, blueberries, strawberries or mixed frozen berries. Purée 2 large bananas, 1 cup (150 g) frozen berries and 1 cup (225 ml) coconut milk in a blender until smooth. Pour into ice bloke moulds. Freeze overnight. Makes about 6.

CHOCNANANUT POPSICLES

Purée 2 large bananas, ¼ cup (75 g) hazelnut spread (or cashew or almond spread or no added sugar peanut butter if liked); 2 tablespoons cocoa powder and 1 cup (225 ml) coconut milk in blender until smooth. Pour into ice block moulds. Freeze overnight. Makes about 6.

CORN ON
A STICK

Cook corn cobs in plenty of water for 6–8 minutes and drain. Skewer onto thick wooden sticks. While hot spread with some butter and finely chopped herbs, such as coriander, parsley, thyme and a sprinkle of paprika.

POPCORN

Put 2 tablespoons popping corn in a large microwave-safe bowl. Toss in ¼ teaspoon each olive oil and lemon pepper (available in specialist shops). Cover the bowl with a double-thick layer of plastic wrap and pierce a couple of times. Cook on high for 3–4 minutes until the popping stops.

THE OCCASIONAL REWARD
PEANUT BUTTER CUPS

½ cup (75 g) roasted peanuts

1 lady finger banana

2 tablespoons melted coconut oil, + ¼ cup (55 g) extra

¼ cup (25 g) raw cacao powder (from health food stores and some supermarkets)

Great for those who still crave a bit of chocolate every once in a while.

Blend together the peanuts, banana and 2 tablespoons of coconut oil in a blender until smooth (don't worry if the peanuts are still a bit chunky).

Using a chocolate mould, mini cupcake wrappers, egg cups or ice cube trays, scoop the peanut mixture, about three-quarters filled into the moulds, place in the freezer to cool rapidly or just refrigerate if you are not in a hurry.

Meanwhile, melt together the extra coconut oil and cacao in a small saucepan over a low heat until smooth. Pour into a small jug.

Once the peanut mixture is firm, pour the chocolate mix over. Freeze again and *voilà!*

Alternative Butter Cups: *Use ½ cup (150 g) no added sugar peanut butter mixed with the 2 tablespoons melted oil in place of the peanuts and banana.*

MAKES ABOUT 12 ICE CUBE TRAYS

(BUT NUMBER DEPENDS ON THE SIZE OF THE CONTAINER YOU USE)

BLUEBERRY CACAO CLUSTERS

3 tablespoons coconut oil

3 tablespoons raw cacao powder
(from health food stores and
some supermarkets)

1 cup (115 g) almonds, lightly
toasted

⅓ cup (55 g) coconut flakes,
lightly toasted

1 punnet fresh or frozen
blueberries

Few fresh violets, optional

An absolute knockout! You can play with the balance of chocolate and nuts too.

Put the coconut oil and cacao in a medium saucepan and stir briefly over low heat until smooth.

While it is still warm but no longer on the heat of the stove add the almonds, coconut flakes and blueberries (if using frozen blueberries add them at the last minute and stir in quickly to stop the mixture clumping).

Spoon lumps of the mixture onto a baking tray lined with baking paper. I recommend ½ tablespoon-sized lumps. Top with a violet, if you have them.

Place in the freezer or refrigerator depending on how soon you want to consume them. Refrigerate for about a day or leave in the freezer for a couple of hours.

Store in an airtight container in the refrigerator for 2–3 days or keep frozen for longer.

Variation: Citrus Cacao Clusters – replace the blueberries with small mandarin segments.

MAKES ABOUT 16

THE ONCE-A-YEAR BIRTHDAY CAKE
RAW FIG AND COCONUT CAKE WITH AN ALMOND CHIA BASE

BASE

2½ cups (310 g) blanched almonds, soaked in water and drained

1 pitted dried date

1 dried fig

½ cup (75 g) chia seeds

½ cup (50 g) shredded or desiccated coconut

½ cup (115 ml) coconut oil

½ cup (115 ml) water

TOP LAYER

2 cups (300 g) cashews, soaked in water and drained

1 pitted dried date

2 dried figs

400 g tin coconut milk

½ cup (50 g) desiccated coconut

Please note that this cake still contains fructose in the form of dried fruit, but it's much healthier for a special occasion than the standard supermarket cake.

Put all the ingredients for the base in a blender or food processor and blend until well combined. Spread the mixture evenly into a cake tin, pressing down until it's compact. Take special care to get the edge even.

Rinse the blender. Then blend all the ingredients for the top layer together until smooth. Pour over the base layer in the cake tin.

Freeze overnight. Decorate after freezing with your choice of edible flowers, fruits or nuts. We used thyme, clover and rosemary flowers. (You can also use violets or other edible flowers.) Before serving the cake place in the refrigerator to soften a little. Cut with a knife that's been run under hot water.

Hints: *Always soak the almonds and cashews in two separate bowls overnight or a minimum of two hours before blending.*

Melt the coconut oil by running the jar under hot water before measuring to make it easier.

Note: *I have been known to throw a slice in the blender to make a delicious smoothie.*

'THROUGHOUT MY RESEARCH I HAVE
CONVINCED MYSELF THAT SUGAR
IS THE PROBLEM AND WITHOUT
SUGAR EVERYTHING ELSE BECOMES
RELATIVELY HARMLESS.'
GARY TAUBES, AUTHOR OF *WHY WE GET FAT*

CONCLUSION/EPILOGUE

The past two years of my life have involved a lot of sugar. I have experienced filling my body to capacity with it; I have dreamt about it every second night; and I have read nearly every sugar-related article or scientific study on the internet. In fact, our three-month-old daughter can consider herself very lucky not to be named Sugar. I can't even bring myself to call her sweetheart.

It actually feels like I was dating sugar all that time. She was so attractive at first, full of energy and playful, but then we moved in together and the truth began to emerge. She did my head in, had very little substance and really inflicted some physical damage. I am happy to accept she wasn't the one for me and I'm ready to move on. That doesn't mean however that she isn't for everyone and I'm not here to tell you who you should or shouldn't date. But please, if you are going to spend some time with her, just make sure it's a quick and heated affair; see her sparingly and on special occasions and whatever you do, do not take her back to your house. She has a habit of getting very comfortable and moving in quickly. She'll sweet-talk your kids – they'll love having her around and before you know it you will be powerless to her charms. I can categorically say that I am a better man without her and although she may still send me a text every now and again, I have learnt not to respond.

If I could summarise my entire experience in a few words it would be: YOU ARE WHAT YOU EAT. I've heard this phrase so many times but now I know what it means at a very deep level. If we all truly understood these words, we would be living on a far healthier planet. Sugar is a quick fix. It is a false energy rush and it affects the way we perceive the world and interact with others. If you want to experience the true depth of relationships, of work satisfaction, of the joyous and healthy life that is available to you, sugar should be off the menu.

This is our one precious moment to live. What if we could halve the time we spend counting calories, sweating in the gym, or popping a variety of pills, just by removing or limiting certain sugars? By educating our kids, we can change the paradigm. Our purpose should be to pass on wisdom to the next generation so they can live happier lives than us. The current view is that this generation will be the first in history to live shorter lives than their parents. That's a tragedy.

Many debates will surface on this topic in the years to come. Studies will emerge from both sides and the food and beverage industries are likely to pull out a few more tricks yet. The thing to remember though is that these industries only have the power because we give it to them. We can choose what we eat. At present, high-sugar foods are the cheaper option, but if we can steer away from them as a collective, the producers will shift to suit our demands. And remember, I spend the same amount now at the supermarket as I did during the experiment because I am buying foods with healthy fats that fill me up. Who knows what savings I may be making on future medical bills.

For now, just try it for yourself. Try being aware of your sugar consumption and lowering it for a while. See if it affects your state of mind, improves your wellbeing or changes you physically. In my experience, removing sugar is not a fad diet or a weight-loss trend – it is a complete life-changer.

I want to emphasise that this is not a book demanding people to quit sugar. Some people may still be able to have sugar in small amounts and that's great. I know I can't – I'll just want more. Also, I've learnt that I'll get the right amount of sugar I need where it occurs naturally, with fibre, in whole foods and vegetables.

To me the detrimental effects of refined sugar and excess fructose were very real. The proof was in my liver and the other changes to my mind and body; the proof is in an increasingly overweight and sick population relying on medication; the proof is in a young boy who still drinks the soft drink that has destroyed his teeth; and the proof is in an Aboriginal community asking to have sugar removed so they can be healthy once again.

In fact, this time the proof may actually be in the pudding – about 10 teaspoons of it.

If you are interested in learning more, please visit the website for the film and book: **www.thatsugarfilm.com**.

To find out how you can help keep the initiatives of the Mai Wiru, in the Aboriginal communities on the APY lands, alive please visit **www.maiwirufoundation.org**. David Gillespie, John Tregenza and myself have set up a charity that pays for nutritionists to go into the community and train the locals, who in turn help others to make healthy choices as they shop. The aim is to be hands-off and simply keep the local people empowered in a project they developed — one that was successfully reducing the community's levels of sugar consumption. Perhaps they can shine a light for other communities.

USEFUL WORDS/TERMS TO REMEMBER

Adipose tissue: another name for body fat.

Candida: a parasite living in the gut that loves to feed off sugar. It is a type of yeast that when out of control can lead to fungal infections.

Cholesterol (large, fluffy LDL): the misunderstood cholesterol that was believed to be bad for years. It is often associated with saturated fat intake. It turns out it is not the LDL cholesterol that does the damage.

Cholesterol (small, dense LDL): the trouble-makers of cholesterol. This is the type that forms the plaque on your artery walls.

Fructokinase: the unregulated enzyme in the liver that allows the fructose to be pulled out of the bloodstream whether we need it or not. This creates the overload and so it gets turned into fat.

Fructose: the mischievous child of the sugar family. This is a sugar found in fruit and one half of sucrose (table sugar) but we have not evolved to deal with the amount we are currently consuming.

Insulin: the master hormone that controls what we do with the foods we eat. It tells the body to either use fuel for energy or to store it. High insulin levels mean we store fat and don't burn it for energy.

Metabolic syndrome: a collection of conditions that we see so regularly now in the general population. They include heart disease (CVD), type 2 diabetes, hypertension and obesity.

NAFLD (Non-alcoholic Fatty Liver Disease): this is when the liver becomes full of fat. In my case it was the result of all the sugar (fructose) I was eating. It now affects 5.5 million Australians.

Refined carbohydrate: this is when whole plants high in carbohydrate are processed and stripped of everything except the high carbohydrate (starch or sugar). The body processes it very quickly (usually because the fibre and nutrients are removed) so insulin spikes very quickly.

Saturated fats: these fats have often been demonised but recent studies show they may actually be beneficial for us. They include butter, coconut products and aged cheese.

Triglycerides: the form in which most fat exists in food and in the body. They are also found in the blood and are what fructose is converted to in the liver. High levels in the blood are becoming a key marker in determining heart disease.

Type 1 diabetes: a condition where the pancreas stops making insulin. As a result the body cannot turn glucose into energy.

Type 2 diabetes: the pancreas makes some insulin but not enough for your body's needs. This is the most common form of diabetes and now kills somebody every 6 seconds.

IMAGE CREDITS

Illustrations by Alice Oehr: pages 9, 21, 24 (right), 30, 36, 51, 57, 62 (bottom), 69, 71 (top), 72, 89, 90–91, 105, 107, 108, 111, 119, 120, 123, 125, 127, 129, 132, 133, 134, 135, 136, 138 (bottom), 147, 149, 150, 160, 230, 234

Images by Shutterstock: pages 4, 10–11, 20 (left), 22, 37, 40, 46, 48, 55, 61(1), 67, 80, 81, 86–87, 92, 93, 94, 113, 114, 118, 130–131 (2 images), 154, 157, 158–159, 231

Movie stills by Madman Entertainment: pages 12 (top), 15, 16, 17, 18, 19, 20, 23, 24 (left), 26, 28, 33, 34, 42, 45, 52, 55, 59, 61(2–4), 62 (top centre), 65, 66, 77, 78, 82, 84, 100, 102, 116, 121, 138 (top), 141, 143, 145 (far right), 148, 155, 228

Photos by John Laurie: pages 10 (right), 12 (bottom), 25, 27, 31, 39, 71 (bottom), 86 (left), 130 (2 images), 144–145, 147 (bottom left), 152, 158 (left), 162–227

Photos by Marieka Walsh: pages 96, 99

ACKNOWLEDGEMENTS

To my beautiful family, Zoe and Velvet.

To the talented and creative minds that made the sugar story happen: Gareth Davies, Rory Williamson, Nick Batzias, Virginia Whitwell, Jane Usher, Suzanne Walker, Seth Larney, Judd Overton, Trisha Garner, Alice Oehr.

For his constant support, David Gillespie.

For their scientific minds and help, Dr Ken Sikaris, Dr Simon Thornley, Dr Kieron Rooney, Gary Taubes.

For their loving support, Miriam McCaffrie, Jeff Gameau, Jill Testrow, Susie Tuckwell, Alan McConnell.

The Sugar Squad: Dr Debbie Herbst, Sharon Johnston, Andy Garlick, John Tregenza and the Mai Wiru organisation.

The people of the APY lands.

Thank you Monica Gameau, Jason Sourasis, Alethea Jones, Rod Tayler, Rory Robertson, Heather Billings, Scott Butterworth, Paul Wiegard, Marieka Walsh, Macks Faulkron, Dr George Kolouris.

The Pan Macmillan team, Ingrid Ohlsson, Paul O'Beirne, Jace Armstrong, Michelle Earl.

For their help in the early days, Charlie Goldsmith, Sarah Wilson, Susie Tuckwell.

Damon Herriman for his Mountain Dew Mouth story at a café.

All the kind people who agreed to be interviewed: David Wolfe, Michael Moss, Kelly Brownell, Kimber Stanhope, Jean-Marc Schwarz, Professor Nick Allen, Eric Stice, Sonja Yokum, Kathleen DesMaisons, Thomas Campbell, Kathleen Page, Dr Edwin Smith, Larry Hammons & family, Leonard Burton, Stephen Fry, Danielle Reed, Julie Menella, Howard Moskowitz, Cristin Kearns, Barry Popkin, Aaron Matheson, Dr John Sievenpiper.

BIBLIOGRAPHY & SUGGESTED READING

BOOKS

Bennett, Connie and Stephen Sinatra. 2006, *Sugar Shock!* Penguin, New York
Gillespie, David. 2008, *Sweet Poison.* Penguin Books Australia, Sydney
Kessler, David A. 2010, *The End of Overeating.* McClelland & Stewart, Toronto
Moss, Michael. 2013, *Salt, Sugar, Fat.* Random House, New York
Taubes, Gary. 2007, *Good Calories, Bad Calories.* Alfred A. Knopf, New York
Taubes, Gary. 2010, *Why We Get Fat.* Alfred A. Knopf, New York
Yudkin, John. 2012, *Pure, White and Deadly.* Penguin, London

SELECTED STUDIES

Bes-Rastrollo M, Schulze MB, Ruiz-Canela M, *et. al.* Financial Conflicts of Interest and Reporting Bias Regarding the Association between Sugar-Sweetened Beverages and Weight Gain: A Systematic Review of Systematic Reviews. PLoS Med. 2013 Dec;10(12):e1001578.

Brimblecombe JK. Characteristics of the community-level diet of Aboriginal people in remote northern Australia. Med J Aust 2013; 198(7):380–384.

Bronwell KD, Warner KE. The Perils of Ignoring History: Big Tobacco Played Dirty and Millions Died. How Similar Is Big Food? Milbank Q. 2009 Mar;87(1):259–294.

Keys A, Taylor HL, Blackburn H, *et. al.* Seven Countries. *Circulation* 1963 Sep;28:381–95.

Lenoir M, Serre F, Cantin L, *et. al.* Intense Sweetness Surpasses Cocaine Reward. PLoS ONE. 2007; 2(8):e698.

Page KA, Chan O, Arora J, *et. al.* Effects of fructose vs glucose on regional cerebral blood flow in brain regions involved with appetite and reward pathways. JAMA 2013; 309(1):63–70.

Pelsser LM, Frankena K, Toorman J, *et. al.* Effects of a restricted elimination diet on the behaviour of children with attention-deficit hyperactivity disorder (INCA study): a randomised controlled trial. *Lancet* 2011; 377:494–503.

Solnick SJ, Hemenway D. The 'Twinkie Defence': the relationship between carbonated non-diet soft drinks and violence perpetration among Boston high school students. Inj Prev doi:10.1136/injuryprev-2011-040117. Article first published online: 24 October 2011.

Stanhope KL, Havel PJ. Fructose consumption: Recent results and their potential implications. Ann N Y Acad Sci. 2010 Mar;1190:15–24.

Yokum S, Stice E. Teen Weight Gain Linked to TV Food Commercials. *Obesity: A Research Journal.* Article first published online: 24 August 2014.

WEBSITES

www.aaronmatheson.com.au
www.bodyologypps.com.au
www.cerealkillersmovie.com
www.davidgillespie.org
www.gameauland.com
www.mariekawalsh.com.au
www.mothernourish.com

www.nofructose.com
www.sharonjohnston.com.au
www.smashthefat.com
www.sugarpolitics.com
www.thatsugarfilm.com
www.trinityfitnesssolutions.com
www.urthlingz.com

INDEX

First published 2015 by Pan Macmillan Australia Pty Limited
1 Market Street, Sydney, New South Wales, Australia 2000

First published in the UK 2015 by Macmillan
an imprint of Pan Macmillan,
a division of Macmillan Publishers Limited
Pan Macmillan, 20 New Wharf Road, London N1 9RR
Basingstoke and Oxford
Associated companies throughout the world
www.panmacmillan.com

ISBN 978-1-4472-9971-4

9 8 7 6 5 4 3 2 1

A CIP catalogue record for this book is available from
the British Library.

Design and typesetting by Trisha Garner, Design Patsy
Illustrations by Alice Oehr
Editing by Ariane Durkin
Index by Frances Paterson, Olive Grove Indexing Services
Photography by John Laurie
Prop and food styling by Caroline Velik
Additional styling by Jamie Humby
Food preparation by Jamie Humby
Recipes by Michelle Earl, Zoe Tuckwell Smith and Sharon Johnston

Printed and bound by Rotolito Lombarda, Italy

Visit **www.panmacmillan.com** to read more about all our books and to
buy them. You will also find features, author interviews and news of any
author events, and you can sign up for e-newsletters so that you're always
first to hear about our new releases.